Armageddon or Evolution?

Advancing the Sociological Imagination

A Series from Paradigm Publishers

Edited by Bernard Phillips and J. David Knottnerus

Goffman Unbound! A New Paradigm for Social Science
By Thomas J. Scheff (2006)

The Invisible Crisis of Contemporary Society: Reconstructing Sociology's Fundamental Assumptions
By Bernard Phillips and Louis C. Johnston (2007)

Understanding Terrorism: Building on the Sociological Imagination
Edited by Bernard Phillips (2007)

Armageddon or Evolution? The Scientific Method and Escalating World Problems
By Bernard Phillips (2008)

Postmodern Cowboy: C. Wright Mills and a New 21st-Century Sociology
By Keith Kerr (2008)

Struggles before Brown*: Early Civil Rights Protests and Their Significance Today*
By Jean Van Delinder (2008)

The Treadmill of Production: Injustice and Unsustainability in the Global Economy
By Kenneth A. Gould, David N. Pellow, and Allan Schnaiberg (2008)

Bureaucratic Culture and Escalating Problems: Advancing the Sociological Imagination
Edited by J. David Knottnerus and Bernard Phillips (2009)

Forthcoming

Ritual as a Missing Link within Sociology: Structural Ritualization Theory and Research
By J. David Knottnerus (2009)

Armageddon or Evolution?

The Scientific Method and Escalating World Problems

Bernard Phillips

Routledge
Taylor & Francis Group
LONDON AND NEW YORK

First published 2009 by Paradigm Publishers

Published 2016 by Routledge
2 Park Square, Milton Park, Abingdon, Oxon OX14 4RN
711 Third Avenue, New York, NY 10017, USA

Routledge is an imprint of the Taylor & Francis Group, an informa business

Library of Congress Cataloging-in-Publication Data

Phillips, Bernard S.
Armageddon or evolution? : the scientific method and escalating world problems / Bernard Phillips.
p. cm. — (Advancing the sociological imagination)
Includes bibliographical references and index.
ISBN 978-1-59451-606-1 (hardcover : alk. paper)
ISBN 978-1-59451-607-8 (paperback : alk. paper)
1. Sociology—Methodology. 2. Science—Methodology. 3. Social problems. 4. Applied sociology. 5. Sociology. I. Title.
HM511.P55 2008
301.01—dc22

2008024386

ISBN 13 : 978-1-59451-606-1 (hbk)
ISBN 13 : 978-1-59451-607-8 (pbk)

Designed and typeset by Straight Creek Bookmakers.

Contents

Preface

Is IT POSSIBLE TO SEE WORLD PROBLEMS as continuing to increase and yet also see a direction for solving them? Can we come to believe that the continued existence of the human race is now at stake yet also come to believe that we can indeed learn to escape from our highly threatening situation? Given what we hear and see on the mass media from one day to the next, can we still retain what Barack Obama called "the audacity of hope"? Is it indeed possible to be both extremely realistic about the current situation of humanity and also optimistic about the possibility that—following Thornton Wilder's *The Skin of Our Teeth*—we can find a way out of escalating world problems?

My own immersion into the knowledge developed throughout the social sciences has convinced me that at this time in history the fate of the human race hangs in the balance. My reference to Armageddon in the title is by no means an effort to hype the book: I genuinely believe that we humans have only a limited time to gain understanding of the forces threatening us and to develop effective means to counter those forces. In *The Invisible Crisis of Contemporary Society* (Phillips and Johnston, 2007) I and my coauthor documented our increasing problems. And since many of them are invisible—by contrast with, say, global warming, threats from terrorist groups, or nuclear proliferation—they are all the more threatening.

To illustrate our escalating yet invisible problems, there is a widening aspirations-fulfillment gap throughout the world, or a growing disparity between what people want and what they are actually able to get. The scientific, industrial, and technological revolutions over the past five centuries have opened up for all of us previously undreamed possibilities, and there are almost no limits to our material and nonmaterial desires. Millionaires want to become multimillionaires, and multimillionaires want to become billionaires, and almost all of us including the extremely wealthy want to achieve ever more of a sense of self-worth, of living in a truly meaningful way, and of having close personal relationships. Given these accelerating aspirations, they are accompanied by increasing failures to fulfill them, and the resulting invisible aspirations-fulfillment gap is working to accelerate world problems.

One such fundamental problem that remains largely invisible is the increasing gap throughout the academic world between the ideal of following the scientific method and the actual practices—whether in the social, biological, or physical sciences—that differ sharply from that ideal. For example, that ideal requires scientists to open up to all knowledge that is relevant to the problems under investigation. My overall assumption is that human behavior is incredibly complex, as illustrated by our escalating unsolved world problems. Yet there are no fewer than 46 distinct Sections of the American Sociological Association (ASA) populated by sociologists who only rarely communicate with those in other Sections. Further, there are 397 specialized topics within sociology that are discussed in the five-volume *Encyclopedia of Sociology* (Borgatta and Montgomery, 2000), and once again we find limited communication across these many fields of knowledge, as indicated by the limited number of cross-references throughout these specialized areas.

This problem of specialization by scientists coupled with limited communication with scientists in other fields is by no means restricted to sociology. We find it equally in the other social sciences: in psychology, anthropology, political science, economics, and history. And we find it as well throughout the biological and physical sciences. One of the repercussions of such extreme specialization with limited communication across these fields is the escalating cost of higher education, far exceeding the rising cost of living. Given a much broader approach to knowledge, far fewer professors would be needed to cover the range of subjects that students study, and the coverage would be more educational for students, since they would learn to integrate their specialized knowledge to a greater degree. This same argument could also be extended to the many highly specialized fields throughout medicine, where accelerating costs also far exceed the rising cost of living. Here, the ideals of medicine also call for the integration of knowledge—just as every part of the human body interacts with every other part—yet what we all encounter is the general lack of broad understanding on the part of medical specialists. And these problems of the costs of specialization with limited communication remain largely invisible, for I've never seen any discussion of them within discussions of the rising costs of education and medical care and what might be done about them. As a result, what we have in the modern world is not the full use of the scientific method but no more than a partial use of that method, which abandons its full potential for solving problems. This is a most serious matter as social problems throughout the world continue to escalate.

Yet such specialization with limited communication throughout the sciences and throughout medicine is no more than an example of the fundamental structure of contemporary societies, and this helps us to understand why scientists behave as they do: despite their scientific ideals they remain victims of their experiences from one moment to the next as members of society. Organizations throughout the world—as we have learned from the social sciences—are "bureaucratic." This means that they are both highly specialized and hierarchical or "stratified." And this also means that communication across specialized areas as well as up and down those hierarchies is quite limited. The most dramatic example in recent times was the failure of the FBI, CIA, and NSA—all part of the same U.S. government—to communicate with one another effectively about the potential for the 9/11 plot before it happened, with an excellent chance that the 9/11 catastrophe never would have happened had they done so. A far less dramatic example is my own recent experience with water collecting on the new roof of my condominium in Longboat Key, Florida. The roofing company CEO along with my condo association manager insisted that it was the "industry standard" to allow water to collect there because the membrane on the roof is thick enough and sufficiently resistant to prevent any leakage through my ceiling. Yet those individuals were behaving much the same as specialists throughout the world, focusing only on the narrow problem of water penetrating the roof. They failed to consider three other problems that are of no concern to such specialists: the development and spread of mold from standing water, the importance of avoiding standing water so as to eliminate breeding grounds for mosquitoes, and the safety problem that such standing water could pose for anyone going on the roof to initiate repairs.

My argument here is that a "bureaucratic or stratified worldview"—which includes both specialization with limited communication and persisting hierarchy or stratification—is so powerful in the way we have all learned to think, feel, and act that it is able to trump ideals as powerful as those governing scientific and medical practice. Such a worldview is so powerful that it shapes the very ways in which we perceive the world and ourselves from one moment to the next. For example, bureaucracy and stratification are oriented outward to our relationships with others, and it is no coincidence that we also have learned to look outward and avoid seeing the noses on our own faces along with our own bodies as we go about the business of living. But when we are dealing with something as broad, powerful, and concrete as a worldview, we are also dealing with a phenomenon that is invisible, just as the water in the goldfish bowl is invisible to the goldfish. We are dealing with our most fundamental assumptions about the nature of reality, given the name "metaphysics" by philosophers. Unfortunately, scientists generally see metaphysics as something to be avoided because "the nature of reality" is so abstract or general as to be extremely far from what they have come to see as central to the scientific method: very concrete procedures for experimentation or observation. Yet in this way they lose a basic understanding of how the scientific method actually works: testing abstract or general ideas or theories against concrete or empirical evidence. And even the extremely abstract ideas containing our metaphysical assumptions must be tested, for it is those assumptions—such as a bureaucratic or stratified worldview—that shape every aspect of the research process.

My references in these few pages to the aspirations-fulfillment gap, a largely invisible problem that is fundamental throughout the book, and is linked to personal and social problems, give the reader a preview of Figure I-1 in the introduction. And my references to that gap together with a highly specialized approach to the scientific method as well as to a bureaucratic or stratified worldview provide the reader with a preview of Figure I-2, anticipating my overall analysis of the sources of that problem. That relatively invisible problem—following the first paragraph of this preface—is much of the basis for my initial statement that "world problems" are "continuing to increase," that "the continued existence of the human race is now at stake," that we should pay serious attention to "what we hear and see on the media from one day to the next," and that we should be "extremely realistic about the current situation of humanity." But the title of this book is not just "Armageddon." I am not in the business of doom and gloom. The title is "Armageddon or Evolution?" What about the rest of what is in that first paragraph? What about the possibility that we can "see a direction for solving" world problems? What about the idea that we can "learn to escape from our highly threatening situation"? What about "the audacity of hope"? What about the possibility that "we can find a way out of escalating world problems" by "The Skin of Our Teeth"?

We might see that first paragraph as illustrating what might be called a "pendulum metaphor or image"—an idea so concrete that we can perceive or visualize it—that illustrates not only Figure I-2 but also an alternative to Figure I-2 that is sketched in Figure I-3. Let us imagine a pendulum swinging to the left side and then to the right side and swinging in ever-widening arcs. Following the ideals of the scientific method, the swing to the left suggests a commitment to solving a given problem, and the swing to the right suggests some progress in understanding and solving that problem. That progress is suggested by the pendulum's swinging farther to the right than previously. And that wider swing to the right yields additional momentum for a wider swing to the left, suggesting greater commitment to the problem. That greater commitment in turn yields a swing still farther to the right and so on with no limit as to how far the pendulum can swing in both directions. The first paragraph of the preface describes a swing very far to the left as well as very far to the right: entertaining the possibility of Armageddon can help us to entertain the possibility of the further evolution of the human race.

Yet that pendulum metaphor can be employed equally to explain the increasing aspirations-fulfillment gap and increasing personal and social problems, as suggested by Figure I-2. Following that figure, the aspirations-fulfillment gap along with personal and social problems will get worse and worse on the right-hand side to the extent that our response to it is to invoke the only approach that we know: our bureaucratic or stratified worldview and scientific method. Thus, increasing problems will yield greater momentum for the very approach that is much of the basis for increasing those problems further. To illustrate, rising costs of medical care have led to a pending 10 percent cut in Medicare payments to physicians. But the result of this is to motivate many physicians with Medicare patients—who are generally needier than others—to give up treating those patients and focus on those who are more well-to-do. This is further illustrated by the spread of "boutique" medical practices, where patients must pay a substantial amount up-front before they can be treated when medical problems develop. The result of this narrow effort to control medical costs—without taking into account the complexity of the situation—is, then, increasing hierarchy or social stratification. In other words, the bureaucratic worldview of politicians has succeeded in yielding worse problems for poor patients, who will now have access to fewer doctors and overcrowded waiting rooms. The fulfillment of those patients' aspirations for excellent medical care moves farther away from their aspirations.

What is essential in moving from Figure I-2 to Figure I-3—taking into account the urgency of such a move as illustrated by Figure I-1—is an alternative to our bureaucratic or stratified scientific method and bureaucratic or stratified worldview. What is equally essential is that we take into account the relatively invisible nature of what is involved, for that prevents us from seeing the enormous problem that we face. To further illustrate these problems, let us consider the audiences for this book. Sociologists and other social scientists, following their bureaucratic or stratified worldview and approach to the scientific method, will look for evidence that I have read and taken into account the highly specialized published articles and books that they have been working with over many years. Failing my ability to demonstrate that—and it would take a book of many thousands of pages to do that adequately for social scientists in general, given their thousands of specialized areas—this book will lose credibility in their eyes. As for the more general readers, to the extent that I make any effort to bring forward the technical developments throughout the social sciences—as illustrated by Figures I-1, I-2, and I-3—they will come to believe that the book is too difficult for them to understand and will lose interest in it. In both cases, the bureaucratic or stratified worldview will succeed in trumping my own efforts to communicate to a wide audience.

Considering this situation, however, my task is by no means hopeless. For one thing, given the increasing problems throughout the contemporary world, there will be greater willingness to question traditional procedures for solving problems and traditional ideas throughout the social sciences. For another thing, I am betting on the incredible potential of every reader and every human being to learn from experience, and our shared experience has been that problems are indeed increasing. On my own part, what I must attempt to do is to follow the pendulum metaphor—in the context of Figure I-3 but not Figure I-2—to demonstrate that problems are indeed increasing despite efforts to solve them. In other words, I must move that pendulum far to the left to help convert invisible problems like the aspirations-fulfillment gap to visible problems. In addition, I must provide an alternative approach to the scientific method that promises to make progress in solving our growing problems. Such an alternative must build on how the scientific method has actually worked effectively over the years, taking into account the scientific ideal of opening up all knowledge relevant to a given problem. Further, and this will be more difficult—given the incredible complexity and invisible nature of a worldview—I must provide an alternative worldview that also promises progress in solving our fundamental problems.

Yet I am not working alone in these efforts. The effectiveness of the physical and biological sciences over five centuries has convinced me of the incredible power of the scientific method to solve fundamental problems, granting that the order of complexity of those problems does not approach the problems of human behavior. The insights from the humanities—from philosophers, novelists, playwrights, and poets—have strengthened my conviction as to the infinite potential of every single human being. And the achievements of social scientists in the nineteenth and twentieth centuries—granting the fact that they have yet to be integrated—are equally encouraging. More specifically, one figure within the history of social science has followed the pendulum metaphor to a substantial degree and has given me a clear direction for building on his work: C. Wright Mills. I was a premedical student at Columbia University in the early 1950s until I chanced to take several of Mills's courses, and that changed my entire life. Mills's books were written—as this one is—not simply for sociologists but for a much broader audience. Following the pendulum metaphor, he was deeply committed to confronting the fundamental problems of the modern world, as is well illustrated by his books *White Collar* (1951), *The Power Elite* (1956), *The Causes of World War Three* (1958), and *Listen, Yankee: The Revolution in Cuba* (1960). At the same time that he focused on problems, he also focused on solutions, both in these books and in a book that was voted by the members of the International Sociological Association as the most influential book for sociologists published during the entire twentieth century: *The Sociological Imagination* (1959).

That book pointed a direction for a much broader approach to the scientific method, one that follows the ideal of opening up to a full range of phenomena that are relevant to any given research problem, as illustrated by this passage:

> The sociological imagination . . . is the capacity to shift from one perspective to another—from the political to the psychological; from examination of a single family to comparative assessment of the national budgets of the world; from the theological school to the military establishment; from considerations of an oil industry to studies of contemporary poetry. It is the capacity to range from the most impersonal and remote transformations to the most intimate features of the human self—and to see the relations between the two. (1959: 7)

Although I had written four textbooks on research methods throughout the latter part of the twentieth century—influenced deeply by Mills's vision of the possibilities of social science—it was only after my retirement from active teaching in 1999 that I became serious about confronting the failure of the social sciences to live up to those possibilities. I started to organize informally what came to be called the Sociological Imagination Group, meeting annually starting in 2000. In my *Beyond Sociology's Tower of Babel: Reconstructing the Scientific Method* (2001), I attempted to demonstrate and illustrate the possibility of not only building on Mills's understanding of the scientific method in a highly systematic way but also building most substantially on the work of two colleagues who had joined me, Harold Kincaid and Thomas J. Scheff. Kincaid, a professor of philosophy, helped me to understand both the limitations of traditional assumptions by social scientists about the scientific method as well as the importance of alternative assumptions that follow scientific ideals. Scheff taught me the enormous complexity of the momentary situation, such as the emotional dynamics that are involved, based in part on three monographs he had written during the 1990s. The three of us proceeded to edit a volume of the papers presented at the first meeting of the Sociological Imagination Group: *Toward a Sociological Imagination: Bridging Specialized Fields* (2002). That approach to the scientific method—which we came to call the Web and Part/Whole Approach—was developed further in subsequent books: *The Invisible Crisis of Contemporary Society* (Phillips and Johnston, 2007), *Understanding Terrorism* (a collection of papers presented at our San Francisco conference in 2004: Phillips, ed., 2007), and *Bureaucratic Culture and Escalating Problems* (a collection of papers presented at our Montreal conference in 2006: J. David Knottnerus and Phillips, eds., forthcoming).

All of these five books—including the first one, *Beyond Sociology's Tower of Babel*—focused not only on a broad approach to the scientific method in order to fulfill scientific ideals but also on our worldview, which is powerful enough to trump scientific practices that fail to conform to it. More specifically, they contrasted a bureaucratic or stratified worldview pointing away from scientific ideals with an interactive or evolutionary worldview that reinforces those ideals. The difficulties involved in this effort are enormous, given the breadth of worldviews, their invisible nature, and the orientation of social scientists in general to define the scientific method mainly in terms of concrete procedures rather than abstract ideas like worldviews, which are metaphysical in nature. Nevertheless, granting the limited attention to worldviews by scientists, it is indeed possible to make progress in understanding their nature. For example, one of the two major conclusions of *The Invisible Crisis of Contemporary Society* (with the other being evidence for an increasing aspirations-fulfillment gap) was this: *"To the degree that a worldview or metaphysical stance is stratified* [or bureaucratic] *versus interactive* [or evolutionary], *there will be a large gap between aspirations and their fulfillment"* (Phillips and Johnston,

2007: 234). This conclusion also implies its converse: To the degree that a worldview or metaphysical stance is interactive or evolutionary, there will be a small gap between aspirations and their fulfillment.

Yet this is no more than a beginning in any effort to understand the nature of these two worldviews or their impact on us. Exactly what is their nature? Although we know a fair amount about bureaucracy, ideas about evolution have varied greatly. Just what is an "evolutionary or interactive worldview"? The idea of evolution has been used in many different ways, and a number of them have hurt rather than helped our understanding of human behavior and have even contributed to our problems. We might think of Hitler's vision of Aryans as the "master race," and his mounting the Holocaust as a way of eliminating "inferiors" such as Jews and supposedly moving toward a higher development of the human race. We might think of Charles Darwin's theory of biological evolution as advancing our understanding of biology to an enormous degree. We might also think of Auguste Comte—the early founder of sociology—with his vision of the human race as necessarily going through three stages: theological, metaphysical, and positive, where the positive stage is the age of science and industrialism. Or there was Herbert Spencer's theory of necessary movement from homogeneity to heterogeneity or specialization. And we have another sociologist, Ferdinand Tonnies, and his view of necessary movement from *Gemeinschaft* or community to *Gesellschaft* or impersonal society, an evolutionary movement that is not in a progressive direction. More recent theories of evolution are illustrated by the anthropologist Leslie White and the sociologists Gerhard and Jean Lenski who emphasize the importance of technological development.

My own approach to the nature of an "evolutionary worldview" has to do not with biological evolution but with the development of the human being's personality structure along with the social structure of human societies. Yet it is similar to Darwin's theory of biological evolution, as explained in part by the following passage from *The Structure of Scientific Revolutions* (1962) by Thomas Kuhn, a most influential historian of science:

> All the well-known pre-Darwinian evolutionary theories ... had taken evolution to be a goal-directed process. The "idea" of man and of the contemporary flora and fauna was thought to have been present from the first creation of life, perhaps in the mind of God. That idea or plan had provided the direction and the guiding force to the entire evolutionary process. Each new stage of evolutionary development was a more perfect realization of a plan that had been present from the start. For many men the abolition of that teleological kind of evolution was the most significant and least palatable of Darwin's suggestions. The *Origin of Species* recognized no goal set either by God or nature.... Even such marvelously adapted organs as the eye and hand of man—organs whose design had previously provided powerful arguments for the existence of a supreme artificer and

> an advance plan—were products of a process that moved steadily from primitive beginnings but toward no goal. (1962: 170–171)

Kuhn distinguishes here between "evolution toward" and "evolution from," with the latter rather than the former corresponding to Darwin's theory of biological evolution. There are no inevitable movements in a progressive direction, no necessary stages that the evolutionary process travels through, and a great deal of variation in the evolution of the multitudes of flora and fauna on the planet, depending on the particular and complex circumstances within each evolutionary situation. For me as well as for Kuhn, there are similarities between this process of biological evolution and our understanding of the nature of the scientific method. Just as biological evolution does not proceed "toward the realization of a plan that had been present from the start," neither does the scientific method—when it is practiced according to our ideals—proceed so as to conform to some hidden agenda, such as a bureaucratic or stratified worldview with its focus on specialization with limited communication and persisting hierarchy or social stratification. Unfortunately, however—due to what I am convinced is the universality of our bureaucratic or stratified worldview—practices throughout the many sciences do indeed follow that hidden agenda. It is an agenda that is well illustrated by the near-universal failure of scientists to measure and report on their own impact on the research process at its every stage, thus closing off the possibility of their own genuine openness to and interaction with their research environment. For example, an interviewer will assess the responses of an interviewee yet fail to assess his or her impact on the interviewee, an impact that will work to color and shape those responses in one way or another. Granting that other researchers do have an opportunity to corroborate a given investigator's findings, they also are subject to the same outward orientation of a bureaucratic or stratified worldview; and as a result they will not uncover this failure to assess that coloring and shaping. They too will remain insensitive to the central importance of what has come to be called "investigator effect."

Movement in the direction of "evolution from" rather than "evolution toward"—and also in the direction of what I have called an "evolutionary worldview"—requires a scientific method that is extremely broad, one that is broad enough to take into account investigator effects. The history of the social sciences has seen the collection of a vast amount of knowledge of human behavior, and this is indeed an incredible resource for us today as we proceed to confront the problems that threaten modern society. Yet that knowledge is located in relatively unconnected bits and pieces, given the prevailing bureaucratic or stratified worldview. Those pieces must be integrated to penetrate the complexity of those problems in any profound way, following the ideals of the scientific method rather than current practices. Looking to Figure I-3, such a scientific method can—a little at a time—yield a shift toward an evolutionary worldview, and also work to narrow the aspirations-fulfillment gap, resulting in some progress in solving personal and world problems.

Given the discussion of the background of this book, how might we proceed? Now that there is evidence for increasing problems—a major conclusion of *The Invisible Crisis of Contemporary Society*—I want to further develop that sense of problem, suggested by the metaphor of the pendulum's moving farther to the left. For many of our fundamental problems, such as the aspirations-fulfillment gap, remain largely invisible. Unless it becomes clear to the reader that we are all facing a mammoth crisis that is continuing to advance, there will be no motivation to confront the most difficult tasks of not only our traditional approach to the scientific method but also our bureaucratic worldview, one that guides all behavior from one moment to the next. It is such a deep sense of problem that is located at the heart of a broad approach to the scientific method. Mills's own deep commitments to confronting world problems were much the basis for his ability to communicate so effectively and so widely not only to sociologists but also to other audiences.

The importance of a deep sense of problem—not just about the failure of the traditional approach to the scientific method but also about the link between our bureaucratic way of life and our escalating problems—cannot be overestimated. Yet that deep sense of problem will quickly disappear unless our pendulum can make use of the momentum for movement far to the right. Throughout my own career I developed many strong criticisms of the traditional approach to the scientific method, and I also suggested directions for a much broader approach that followed scientific ideals, an approach not far from that outlined in my *Beyond Sociology's Tower of Babel.* Yet I never was able to venture as far as to criticize our bureaucratic worldview and way of life. Part of my problem—shared by most social scientists and based on our invisible commitment to a bureaucratic worldview—was a commitment to a specialized view of the social scientist by contrast with those who apply scientific knowledge, such as political leaders, teachers, social workers, psychotherapists, nurses, and business people. Just as physical scientists leave to engineers the business of applying physical science to solve problems, so too did I implicitly see myself as developing basic knowledge while leaving to others the business of using that knowledge to solve problems. As a result, my pendulum was limited in its movement to the left, and this limited its movement to the right.

As for movement far to the right, perhaps the most important finding of the social sciences throughout the entire twentieth century is the overwhelming importance of language in shaping human behavior. It is language more than any other characteristic of human beings that distinguishes us from all other forms of life, granting that some mammals have been able to develop the very beginnings of language. In my recent books, I concluded that we humans have made only limited usage of the three potentials of language. First is the gradational orientation, as illustrated by the quantitative and numerical emphasis throughout the physical as well as much of the biological sciences. Then there is the dichotomous, either-or emphasis within our ordinary usage as well as throughout the social sciences, for every word we use divides the world in two: that to which the word refers, on the one hand, and everything else, on the other hand. Third is language's orientation to the senses, imagery, metaphors, or figures of speech, as emphasized throughout the humanities. Just as we have specialization among these three broad fields of language, so do we all learn to limit our usage of all three potentials of language. Further, dichotomy is far more predominant than gradation or metaphor—regardless of our profession—because that is the emphasis of everyday speech and thought.

Alvin Gouldner, another sociologist whom I knew personally, developed a vision of how all of us can move toward making fuller use of the infinite potential of language:

> The pursuit of ... understanding, however, cannot promise that men as we now find them, with their everyday language and understanding, will always be capable of further understanding and of liberating themselves. At decisive points the ordinary language and conventional understandings fail and must be transcended. It is essentially the task of the social sciences, more generally, to create new and "extraordinary" languages, to help men learn to speak them, and to mediate between the deficient understandings of ordinary language and the different and liberating perspectives of the extraordinary languages of social theory.... To say social theorists are concept-creators means that they are not merely in the knowledge-creating business, but also in the language-reform and language-creating business. In other words, they are from the beginning involved in creating a new culture. (Gouldner, 1972: 16)

In this passage Gouldner helps us understand how to move our pendulum very far to the right where we learn to solve problems. Gouldner was no match for Mills in conveying to his readers a very deep sense of problem. But he was able to put forward ideas that advance the development of the kind of broad scientific method that Mills called for in *The Sociological Imagination* (1959). Just as physical scientists make full use of the technical language of physics—with concepts such as "force," "atom," and "valence"—so must those of us who are struggling with understanding the incredible complexity of human behavior make full use of the technical language of the social sciences. That language is illustrated by the concepts of "bureaucratic" and "stratified" introduced in this preface and that will be used throughout the book. Such concepts do not oppose what ordinary language has taught us. Rather, they help us further the insights that we gain from ordinary language. For Gouldner, language is not just an effective way of thinking and communicating but also a way of solving our problems, including the most difficult ones that scientists confront.

This book represents my effort to build, not only on the contributions of Mills and Gouldner, but also on the five books linked to the Sociological Imagination Group that were cited earlier. As a result, this book focuses on the relatively invisible problem of the aspirations-fulfillment gap, given its close links to a wide range of fundamental problems throughout contemporary society. Further, it distinguishes between a bureaucratic or stratified scientific method and worldview versus an evolutionary or interactive scientific method and worldview. Pointing toward an applied direction—by contrast with the neutral emphasis throughout the social sciences—this book focuses on how to move from the former to the latter scientific method and worldview. This book also makes use of language's dichotomous, gradational, and metaphorical potentials, all within the context of putting forward what Gouldner called for: the "extraordinary" language of the social sciences. Overall, this book is guided by the pendulum metaphor, moving as far to the left and right sides as possible.

I am aware that the books written within the context of the Sociological Imagination Group have made only a partial case for the failures of a traditional approach to the scientific method along with a link between our bureaucratic or stratified worldview and escalating world problems. I am also aware that these books have made only a partial case for the potential of the Web and Part/Whole Approach to the scientific method coupled with an evolutionary or interactive worldview for confronting those problems effectively. And I suspect that this book will do no more than make a partial case for moving from the former approaches to the scientific method and worldview to the latter ones. To make a credible case to the full range of social scientists would require a book that would dig deeply into hundreds of literatures throughout the social sciences. Such a book would be thousands of pages long and would put off readers who are not professional social scientists. It would be a book that I—given the limitations of my own background—would be unable to write.

Instead, I have written a book that is suggestive far more than conclusive. Yet it is my hope that it will be sufficiently suggestive that social scientists—granting the book's limited analyses of literatures throughout the social sciences—will become motivated to learn more about the publications and activities of the Sociological Imagination Group. To that end, our Web site—www.sociological-imagination.org—provides excerpts from our books, other material, and information about our annual conferences. And I would welcome correspondence to my e-mail address: bernieflps@aol.com. Given what I am convinced are escalating social problems, I hope that social scientists will join me and others in the Sociological Imagination Group in our efforts to move as rapidly as possible toward a scientific method that follows scientific ideals and toward a worldview that helps all of us confront those problems ever more effectively.

To put the matter more forcefully, I am convinced—based on both the publications cited earlier and my own personal experiences—that the human race is heading rapidly toward extinction, given growing visible and invisible problems. I am also convinced that it is social scientists—far more than any other group—who hold the keys that can unlock the door to understanding our complex and threatening problems. Further, I am equally convinced that the present direction of the social sciences—whether in our teaching, our research, our publications, our conferences, or our communications—is making little headway relative to those problems. Whether the previously mentioned publications and this one yield a direction for making more headway on those problems, I am also convinced that social scientists throughout the world must act as quickly as possible to take responsibility for their crucial role in learning how to understand and solve those problems. In this book I am attempting to communicate these convictions and make them credible to readers. If social scientists share these convictions yet fail to act on them, then it is they who will share a guilt—far greater than that of the Nazis for the Holocaust—for the end of humankind.

As for the audience of general readers who are not social scientists, I believe that this book can indeed be understood by the general reader, given its limited use of technical language and my own efforts to explain the technical concepts I do employ. Here, the Web site previously mentioned with its short papers can help the general reader. Two short papers are "Evolutionary Manifesto" and "Manifesto 2." A longer paper that is not particularly technical—dealing with problems encountered by the residents of nursing homes—is "Institutionalized Elder Abuse." There is also the first draft of an introductory chapter to a new book that I and a colleague are working on: "Manifesto for Deep Democracy." Also included is a short biography of C. Wright Mills that I published. It is my hope that general readers will become interested in making use of sociological ideas in their efforts to confront personal and world problems.

The crucial problem that you will face is a bureaucratic worldview teaching you that only the experts or professionals can solve such problems. Yet our growing problems throughout the world suggest the severe limitations of our experts. Given this situation, I believe that you, as general readers, should consider carefully your own personal responsibilities relative to the world situation at this time in history. If the experts generally have failed us, and if they are continuing to fail us, what kind of future can we expect for ourselves and for our children and grandchildren? And if you have a capacity no less than that of our experts to learn and confront the world's deepening problems—in addition to your own personal problems—what are you doing about it? Are you simply marking time, waiting for the bombs to fall? Or are you proceeding to learn how to make fuller use of your human capacities to act decisively and effectively in the face of threatening problems?

I would like to acknowledge the substantial contributions that J. David Knottnerus—my coeditor of the "Advancing the Sociological Imagination" series with Paradigm Publishers has made to my ideas in general and to this book in particular. His theory of structural ritualization has shaped my own understanding of the rituals of individuals and groups, and his insights with respect to this manuscript have helped me to revise it. I am also deeply appreciative of David Christner's faith in my own possibilities and his encouragement to write a book for readers who are not only sociologists. And I want to add my feelings of gratitude to Dean Birkenkamp, a publisher and a friend whose encouragement has been no less important to me, and whose vision for what should be published at this time in history sets a high standard for publishers throughout the world.

My own overall direction for this book and the others I've written within the context of the Sociological Imagination Group is based largely on the work of two colleagues, Harold Kincaid and Thomas J. Scheff, and I am most grateful to them. Louis Johnston, coauthor of *The Invisible Crisis of Contemporary Society,* has continued over the years to educate me, and his second career as a sociologist following a medical career provides an example for others who have not yet understood the problem-solving potential of the social sciences.

I also want to thank these individuals who have contributed recently to my understanding: Hans Bakker, Stu Bennett, Jerome Braun, Dianne Davis, Frank Elwell, Uta Gerhardt, Elizabeth Gill, Jeff Goldfarb, Ken Gould, Kevin Gotham, Douglas Hartmann, Paul Johnson, Debbie Kasper, Joan Kennedy, Keith Kerr, Louis Kontos, Douglas Meyer, Vince Montes, David Pellow, Michael Phillips, Todd Powell-Williams, Adam Rafalovich, Salomon Rettig, Jim Roach, Yaffa Schlesinger, Allan Schnaiberg, Sandro Segre, Arlene Stein, David Stearns, Emek Tanay, Alex Thornburg, Jonathan Turner, Jason Ulsperger, and Jean Van Delinder.

Finally, there are about a hundred individuals I mention near the beginning of the introduction—sociologists, other social scientists, philosophers, novelists, educators, and other applied scientists—whose ideas have set the stage for this book. To the extent that this book amounts to anything, it is because of a broad framework that reaches out to the works of those authors. Each one of them illustrates the incredible potential of the scientific method, and my thanks go out to them for their contributions.

Bernard Phillips
Longboat Key, Florida
January 2008

PART I

Introduction

I would like to recommend that readers begin with the preface before reading this introduction. By so doing, they will see more clearly my rationale for the approach that I am taking to this book. Further, the preface and introduction reinforce one another in their complementary ways of emphasizing accelerating world problems and giving directions for solving them. That reinforcement is urgently needed, given what I see as a near-universal failure to understand the threats posed by the present world situation to the future of the human race.

And I would also like to alert the reader to my own problem of attempting to reach out to—and gain credibility with—three different audiences: sociologists, other social scientists, and general readers. My own approach to knowledge is as broad as I can make it, given my conviction as to the incredible complexity of human behavior in comparison with physical and biological phenomena. Yet the high degree of specialization coupled with limited communication to be found both inside and outside the academic world works against efforts to integrate knowledge. For example, within sociology alone are some 400 highly specialized areas, and specialists in any given area look for the analysis of studies that they believe are essential for making progress in their particular fields of knowledge.

Given this situation, which I describe in the preface, the best that I can do is to alert the reader to this problem along with my own limitations. What may help to some extent is that the arguments in this book are closely supported by five other recently published books: *Beyond Sociology's Tower of Babel: Reconstructing the Scientific Method* (Phillips, 2001), *Toward a Sociological Imagination: Bridging Specialized Fields* (Phillips, Kincaid, and Scheff, eds., 2002), *The Invisible Crisis of Contemporary Society: Reconstructing Sociology's Fundamental Assumptions* (Phillips and

Johnston, 2007), *Understanding Terrorism: Building on the Sociological Imagination* (Phillips, ed., 2007), and *Bureaucratic Culture and Escalating Problems* (Knottnerus and Phillips, eds., forthcoming). Those books built on the work of a wide range of authors from diverse fields.

The emphasis of those works has been on the literature of sociology, linking their arguments to work by these authors, and not just with quick references: Andrew Abbott, Jeffrey Alexander, Hans Bakker, Howard Becker, Joseph Bensman, Albert Bergesen, Donald Black, Herbert Blumer, Pierre Bourdieu, David Britt, Walter Buckley, Lawrence Busch, Charles Derber, Emile Durkheim, Frank Elwell, Eric Erikson, Richard Farson, Anthony Giddens, Erving Goffman, Alvin Gouldner, Antonio Gramsci, Jurgen Habermas, Douglas Hartmann, Chanoch Jacobsen, Debbie Kasper, James Kimberly, David Knotttnerus, Louis Kontos, Richard Lachmann, Marc LaFountain, Jack Levin, Omar Lizardo, Nicholas Luhmann, George Lundberg, David Maines, Karl Marx, Robert Merton, Robert Michels, C. Wright Mills, David Moberg, Vince Montes, Anthony Oberschall, William Ogburn, Fritz Pappenheim, Suzanne Retzinger, Thomas Scheff, Sandro Segre, Georg Simmel, David Snow, Arlene Stein, Piotr Sztompka, Charles Tilly, Jason Ulsperger, Arthur Vidich, Immanuel Wallerstein, Max Weber, Robin Williams, and Erik Olin Wright.

Other social scientists have not been neglected by those five books that set the stage for this one, as illustrated by substantial links to the works of these authors: Jane Banks, Joan Bondurant, Daniel Boorstin, Clarence Brinton, Donald Brown, Amy Chua, James Davies, Leon Festinger, Ted Gurr, Karen Horney, Susanna Hornig, George Kelly, Thomas Kuhn, Daniel Lerner, Leo Lowenthal, Walter Ong, Milton Rokeach, Robert Rosenthal, and Robert Sommer.

As for the general reader, those five books also included substantial links to work by these philosophers, novelists, educators, and other applied individuals: Edwin Abbott, John Berger, Robert Browning, John Dewey, Paulo Freire, Mohandas Gandhi, Hermann Hesse, Fred Hoyle, Ivan Illich, Harold Kincaid, Alfred Korzybski, Friedrich Nietzsche, George Orwell, Carlos Pecotche, Charles Peirce, W. V. O. Quine, J. S. Ullian, Jack Vance, and A. E. van Vogt.

These efforts to build on and integrate existing knowledge, however, do no more than scratch the surface of available yet unintegrated knowledge bearing on the broad problems that I am investigating. Given what I am convinced are increasing world problems posing a most threatening situation for humanity at this time in history, what is urgently required are widespread commitments by growing numbers of sociologists, of other social scientists, and of the general public to confront these problems. To that end, I hope that this book coupled with the preceding publications of the Sociological Imagination Group—as introduced on the Web site www.sociological-imagination.org—will prove to be useful.

I believe that it is social scientists more than any other group who hold the keys to understanding the deepening problems of the contemporary world. I also believe that we social scientists must learn to take this responsibility very seriously. For this to occur, however, the enormous danger of the present world situation must be communicated effectively to social scientists. In addition, some direction—however preliminary it is and however inadequate it may ultimately prove to be—that promises to make progress on those problems must also be communicated to social scientists, and communicated most effectively. That combination of the problems we face along with a possible solution is suggested by the title of this book: "Armageddon or Evolution?" The importance of this combination is also suggested by the pendulum metaphor discussed in the preface. For this combination can point us toward ever deeper commitment to those problems and ever more progress in understanding them, using a broad approach to the scientific method that succeeds in challenging our bureaucratic or stratified worldview and points us humans in an evolutionary direction.

Chicken Little claimed that the sky is falling, and he may not have been exaggerating. World War II with its massive annihilation of civilian populations and its Holocaust, the cold war with its threats of nuclear devastation, the mounting evidence for global warming and its potential for catastrophic consequences, and the rise of Al Qaeda coupled with the events of 9/11 are only some indications of the increasing problems that confront us and threaten our very survival. For the first time in any era we have weapons capable of destroying all human beings on earth, yet we have little understanding of how to prevent those weapons from being used or ever more destructive weapons from being developed. Never before in history have we been confronted with so many fundamental problems that remain unsolved. Some would argue that the balance of terror between nuclear nations prevents such weapons from ever being launched because nations would fear retaliation. Yet small groups like Al Qaeda would not be deterred by any such fear. Given the continuing development of technologies for producing weapons of mass destruction coupled with the spread throughout the world of those instruments of death, it appears to be just a matter of time before we begin to experience the unthinkable.

Martin Rees, England's Astronomer Royal, has argued in *Our Final Hour: A Scientist's Warning* that "The 'downside' from twenty-first century technology could be graver and more intractable than the threat of nuclear devastation that we have faced for decades" (2003: vii). A small group or even a single individual could learn to concoct a biological cocktail deadly enough to kill many millions or even billions. Is this no more than fear-mongering by "prophets of doom and gloom" to advance their own political agenda? Yet evidence from the work of Rees and others suggests otherwise. The physical and biological sciences have yielded knowledge that has been basic to the shaping of the modern world. Unfortunately, however, that shaping includes the construction of ever more powerful means for destroying that world.

What is happening to us humans, given all of our knowledge, our ability to combat illness and disability, and our achievement in putting a man on the moon? Why are these threats accelerating at this particular time in history? George Lundberg, a sociologist who was much concerned with the failure of sociologists to confront world problems effectively, quotes from an article by Nikita S. Khrushchev in the October 1959 issue of *Foreign Affairs:*

> Is it possible that when mankind has advanced to a plane where it has proved capable of the greatest discoveries and of making its first steps into outer space, it should not be able to use the colossal achievements of its genius for the establishment of a stable peace, for the good of man, rather than for the preparation of another war and for the destruction of all that has been created by its labor over many millenniums? Reason refuses to believe this. (1947/1961: 133)

Back in 1711 Alexander Pope wrote in *An Essay on Criticism* that "a little learning is a dangerous thing." We might see our understanding of physical and biological phenomena coupled with our extremely limited understanding of human behavior as illustrating what Pope called "a little learning." Human phenomena are far more complex than physical and even biological phenomena. We can thus understand the efforts of physical and biological scientists to avoid that complexity in the interest of advancing their understanding of simpler phenomena. We might, then, see such one-sided learning as a fundamental basis for the dangerous world that we are now experiencing. Our knowledge has yielded the AK-47 that schoolchildren might use on their classmates. But it has failed to yield the understanding that would prevent such weapons from being employed. Our "little learning" has yielded a thing-oriented materialistic civilization, since that is what physical technologies can produce. But it has failed to yield the understanding called for by our democratic ideals. For democracy to work we require an education in the complexities of modern problems so that society as a whole can make intelligent political decisions. Yet our limited knowledge of human behavior stands in the way of fulfilling those ideals.

Our situation seems to be similar to that of the passengers who flew United Flight 93 on 9/11. It appears that we too are flying, as we move through space on the planet Earth, on a suicide mission pointing us all toward disaster. Just as most of those passengers in the film on that tragedy probably were completely occupied with decisions like whether to have a Coca-Cola or a Sprite, we too are playing our own games of "trivial pursuit" while failing to gain awareness of our dangerous situation. Sigmund Freud had a good deal to say about such ostrichlike behavior, where we bury our heads in the sand rather than raise our problems to the surface so that we can confront them. This process of avoiding paying attention to problems that threaten the very fabric of our existence has some merit in the short run. It helps us to cope with immediate problems in our everyday lives and achieve some degree of satisfaction as a result, all the while avoiding problems that we are presently unable to solve. In the long run, however, such avoidance works to ensure ultimate disaster.

Yet some of the passengers on United 93 proceeded to examine just what the hijackers were about—as portrayed in the film—and learned the nature of their mission as the plane reversed its direction and headed toward Washington. They developed a strategy that included battering down the cockpit door and moving a passenger with some flying experience into the cockpit. Then, with that memorable phrase transmitted by a cell phone—"Let's roll!"—they proceeded to overpower the hijackers and prevent them from crashing the plane into the White House or the Capitol. Sadly, they were unable to prevent the hijackers from crashing the plane from its low altitude into a Pennsylvania field, killing everyone on board. Can we too learn the nature of our own situation, which evidence suggests is heading us toward destruction? And, paralleling the accomplishment of those few passengers on United 93, can we too develop a strategy for confronting our basic problems? Further, can we parallel the motivation of those passengers, as indicated by their "Let's roll!" commitment, and initiate decisive actions before it is too late to solve our problems?

The Scientific Method

If those passengers were able to learn about their situation and confront their life-threatening problem, the rest of us can learn to do the same. We humans are the product of some four billion years of biological evolution. As a result, we have developed a capacity to learn, using the tool of language, that is far superior to that of any other creature on earth. Thus, we should have the capacity to learn the nature of our own situation. And we certainly have the ability to become motivated to act decisively when our lives are at stake. By far the most important tool that we have for learning, in addition to language, is the scientific method. That is a process that has enabled us humans to shape the world over the last five centuries with a continuing scientific and technological revolution. And that process depends on written language. For the investigator must become aware of the results of previous research relating to the problem at hand to make progress in understanding that problem.

Yet we should allow for the enormous complexity of any social problem whatsoever. Our long-term human history shapes every single social problem that affects us. By contrast, although physical phenomena are indeed affected by prior events, they are rarely changed in any fundamental way, so that simple mathematical formulae can predict their behavior accurately. However, social scientists generally have failed to take into account the full complexity of human behavior as they have attempted to use the scientific method in their research. For example, there are no fewer than 46 distinct Sections of the American Sociological Association (ASA), with each one focusing on a distinct aspect of human behavior. Yet sociologists only rarely communicate with others in different Sections—let alone with others in different social sciences—violating the scientific ideal of taking into account all factors relevant to a given problem. Also, there is no question that social scientists have a substantial impact on the research process, just as interviewers generally influence the responses of those whom they interview. But within the millions of articles and books that have ever been published throughout the social sciences, very few of them have ever examined systematically the impact of what has been called "investigator effect," "experimenter effect," or "interviewer effect."

Edna St. Vincent Millay, writing her poetry early in the twentieth century, had this to say about the failure to integrate knowledge:

> Upon this gifted age, in its dark hour
> Rains from the sky, a meteoric shower
> Of facts . . .
> They lie unquestioned, uncombined.
> Wisdom enough to leech us of our ill
> Is daily spun, but there exists no loom
> To weave it into fabric.
> (Quoted in Phillips and Johnston, 2007: 1)

Just as some of the passengers on United 93 were able to learn about their situation and confront it, so have some sociologists done the same. In particular, it was C. Wright Mills in *The Sociological Imagination* who pointed out a way to respond to Edna St. Vincent Millay's implicit plea:

> The sociological imagination . . . is the capacity to shift from one perspective to another—from the political to the psychological; from examination of a single family to comparative assessment of the national budgets of the world; from the theological school to the military establishment; from considerations of an oil industry to studies of contemporary poetry. It is the capacity to range from the most impersonal and remote transformations to the most intimate features of the human self—and to see the relations between the two. (Mills, 1959: 7)

It was Mills's work that set the stage for the development of a scientific method broad enough to address the full complexity of human behavior, versus the near-universal simplistic approach of specialization with limited communication among specialists throughout the social sciences. Let us note the diversity of what Mills refers to in this paragraph: political and psychological phenomena, a specific family, national budgets, theological schools, military establishments, the oil industry, contemporary poetry, and the individual. To build on Mills's breadth, we might ask how we might approach all of these phenomena—indeed, every single instance of human behavior—in a systematic way.

In 2000 the Sociological Imagination Group was formed to use Mills's ideas about the scientific method as a direction for integrating social science knowledge, an approach outlined in *Beyond Sociology's Tower of Babel: Reconstructing the Scientific Method* (Phillips, 2001; see www.sociological-imagination.org). This required a systematic approach to the language of the social sciences coupled with the knowledge resulting from research throughout those sciences. It is illustrated in ordinary speech by seeing the relationship between general concepts and specific concepts—such as "flower" and "rose"—joined with whatever we have learned about flowers from personal experiences and research. By moving back and forth between general and specific concepts, we come to see the relationship among different kinds of flowers, and we can then add to our knowledge of any given flower by invoking our knowledge of other flowers. Of course, this is the solution to only a very small problem, namely, the problem of understanding relationships among different kinds of flowers. But when we shift from ordinary language to the language of the social sciences, we are able to deal with very large problems.

For example, there is no concept more central to the discipline of sociology than "social stratification" or persisting hierarchy. Its importance is illustrated by the vital role of persisting patterns of hierarchy in work settings that limit opportunities for workers to feel that they are being treated as worthwhile human beings. Such patterns of social stratification are also manifested in a number of other ways, such as patterns of sexism, nationalism, racism, ethnocentrism, religious prejudice, and classism, all of which set up persisting hierarchies. Specialization throughout the social sciences does indeed yield knowledge of particular phenomena. Yet at the same time it works to separate such particular knowledge from the knowledge gained in the other areas. It is as if we viewed roses, daffodils, and daisies as completely different entities without seeing them all as flowers. This violates Mills's advice that social scientists should develop a broad "sociological imagination." It also violates the scientific ideal of reaching out to all knowledge relevant to a given problem.

The systematic approach to the scientific method developed in *Beyond Sociology's Tower of Babel* includes more than Mills's idea of reaching out widely to the range of phenomena relevant to a given problem. In addition, there is Mills's emphasis on the fundamental problems of contemporary society with the aid of such key concepts as "social stratification." Further, it includes Mills's advice that the sociologist pay attention to the relationships between one's personal life and one's professional activities. If researchers are to reach out to the wide range of human phenomena within social science research, then surely we must take into account the role of the researcher as well. This "reflexive" idea, developed more systematically in *Beyond Sociology's Tower of Babel,* was stressed by the sociological theorist Alvin Gouldner in *The Coming Crisis of Western Sociology* (1970):

> What sociologists now most require from a Reflexive Sociology, however, is not just one more specialization, not just another topic for panel meetings

> at professional conventions.... The historical mission of a Reflexive Sociology as I conceive it, however, would be to transform the sociologist, to penetrate deeply into his daily life and work, enriching them with new sensitivities, and to raise the sociologist's self-awareness to a new historical level. (Gouldner, 1970: 487)

Yet how are we to explain why social scientists have failed to follow scientific ideals that require paying attention to all phenomena relevant to a given problem, and that also require assessing their own impact on the research process? If we continue to move toward ever more general linguistic categories, we ultimately encounter our most fundamental assumptions, namely, those that shape to a large extent all of our behavior. Philosophers have a word for this: "metaphysics" or "worldview." Whether we are aware of it or not, all of us make metaphysical assumptions that largely shape our behavior, assumptions such as that we humans have the ability to learn and to solve problems, that we humans have ultimate worth as individuals, and that our relationships with others are of great importance. Those metaphysical assumptions are sufficiently all-encompassing and powerful so as to trump an orientation to follow scientific ideals in a situation where the two are in conflict. Gouldner understood the importance and comprehensiveness of metaphysical assumptions or worldviews, claiming that "these are beliefs about the world that are so general that they may, in principle, be applied to any subject matter without restriction" (1970: 30). It becomes essential, then, to examine the nature of our metaphysical assumptions and to determine whether or not they stand in the way of our scientific ideals and thus are working to prevent the development of social science knowledge.

The Invisible Crisis of Contemporary Society: Reconstructing Sociology's Fundamental Assumptions (Phillips and Johnston, 2007) has focused on those fundamental assumptions or that worldview that guides not only the social scientist's ability to follow scientific ideals but also the behavior of the rest of us from one moment to the next in our everyday lives. The book is based on ideas put forward by Mills, Gouldner, and many others such as those of Thomas Kuhn, the historian of science. Kuhn's *The Structure of Scientific Revolutions* (1962) examined just how scientific revolutions occurred, such as the change from Newtonian physics to Einstein's theory of relativity. He found that scientific revolutions were achieved with great difficulty, requiring the development of new scientific assumptions or "paradigms" that promised to resolve the contradictions within the old paradigms. The problem of changing not just a scientific paradigm but an entire worldview, metaphysical stance, or cultural paradigm should prove to be a far more difficult one; yet the principles involved appear to be much the same.

It is a "stratified or bureaucratic worldview"—with its focus on a narrow orientation to human behavior—that appears to be a basis for the failure of social scientists to follow scientific ideals. Yet given the enormous breadth and relative invisibility of a worldview, how are we to come to grips with it? How are we to learn its nature? What procedures can enable us to understand its impact? Where can we look to develop an understanding of an alternative worldview that would promise to encourage social scientists to follow scientific ideals and also to develop understanding of increasing problems throughout contemporary society?

These questions were addressed in *The Invisible Crisis of Contemporary Society.* Granting the incredible breadth and invisibility of a stratified worldview, it was indeed possible to link that worldview to a problem not quite so broad and not quite so invisible. This was the very same problem that was the focus of my earlier book *Beyond Sociology's Tower of Babel.* Let us, then, proceed to examine the nature of this problem, which is linked to a stratified worldview. By so doing, we should also be able to learn about an alternative worldview that is broad enough to encourage the fulfillment of scientific ideals, one that might be called an "evolutionary worldview." And by so doing we should forget neither the importance of our broad approach to the scientific method nor the importance of learning how to understand the threatening problems of our era.

The Aspirations-Fulfillment Gap

At this time in history we are faced with what appears to be a very large gap between what we want and are able to get. We want a world at peace and we are getting a world full of wars and violence. We want a world where knowledge of physical and biological phenomena continues to develop, yet we are getting a world where that knowledge includes the basis for developing weapons with ever-increasing power to destroy us. We want a world where we understand more and more about human problems, yet the world we are producing includes little understanding of them. More generally, we are constructing a growing gap between our aspirations and the fulfillment of those aspirations. And we might learn to see that dangerous aspirations-fulfillment gap throughout contemporary society. It is this very gap that appears to be tied to a stratified worldview, following the conclusions of *The Invisible Crisis of Contemporary Society.* Correspondingly, it is the narrowing of this gap that appears to be tied to an evolutionary worldview, again following those conclusions.

Following the argument of Emile Durkheim, a key founder of the discipline of sociology, we can see this gap associated with the industrial revolution:

> Such is the source of the excitement predominating in this part of society [trade and industry] and which has thence extended to the other parts. There, the state of crisis and anomy is constant and, so to speak, normal. From top to bottom of the ladder, greed is aroused without knowing

where to find ultimate foothold. Nothing can calm it, since its goal is far beyond all it can attain. (Durkheim, 1897/1951: 254–255)

Durkheim emphasized here the gap between materialistic aspirations and their fulfillment, a gap associated with the scientific and industrial revolutions, for "greed is aroused without knowing where to find ultimate foothold." His concept of "anomy" or "anomie," emphasizing the rapid disorganization of society where norms or widely shared expectations can no longer be counted on, has proven to be a central idea within contemporary sociology. Durkheim also suggested that increased aspirations have been "extended to other parts" of society. Updating Durkheim's view of the extension of this gap throughout society, Harlan Cleveland—a U.S. governmental official—coined the phrase "revolution of rising expectations" in 1950. We have already examined such a gap outside the economic institution in relation to the disparity between our democratic educational ideals and our ability to live up to those ideals. Our scientific and industrial revolutions were accompanied by political revolutions, such as the American and French revolutions, which emphasized a "revolution of rising expectations" in nonmaterial aspects of life. For example, the civil rights movement in the 1960s, the women's movement in the 1970s, and the gay and lesbian movement at the present time all stress the gap between ideals for equality and the fulfillment of those ideals.

It appears that this aspirations-fulfillment gap is central to the fundamental problems of modern society. The U.S. ideal of establishing democracies in Afghanistan and Iraq is to be found alongside mounting deaths and increasing violence throughout the Middle East. The environmental ideal of reversing global warming has been accompanied by limited progress and the strong possibility that such warming will soon prove to be irreversible. Ideals for economic equality have developed throughout the world, yet we are simultaneously faced with an increasing and worldwide economic gap between the rich and the poor. The wide difference between our understanding of biophysical phenomena and human phenomena discussed at the beginning of this section suggests another aspirations-fulfillment gap: between the ideals of social scientists for understanding human behavior and their actual achievements. More generally, we might see this aspirations-fulfillment gap throughout the entire history of the efforts of human beings to solve their problems, efforts that often have resulted in "unanticipated consequences" that have yielded the reverse of initial aspirations. The results of the wars in Iraq and Afghanistan are only recent examples of this phenomenon, which can be illustrated with literally hundreds of historical occurrences (Raushenbush, 1969). One of these examples is the harsh Treaty of Versailles after World War I or "the war to end all wars," which worked to impoverish Germany and also played a central role in the rise of Hitler, World War II, and the Holocaust.

I focused on this aspirations-fulfillment gap in *Beyond Sociology's Tower of Babel: Reconstructing the Scientific Method* (2001). Although my emphasis was on the failure of sociologists to fulfill their scientific ideals through their specialization with limited communication among specialists, I also attempted to explore the origins, general nature, and growth of this gap, including some of its manifestations in the modern world. The story of the chief and the grocer of Balgat and the modernization of that village outside Ankara, Turkey, in the early 1950s illustrates what I learned. Tosun B., an interviewer for Daniel Lerner's cross-national study of modernization (1958), asked the village chief, a 63-year-old man, how satisfied he was with life. He responded: "What could be asked more? God has brought me to this mature age without much pain, has given me sons and daughters, has put me at the head of my village, and has given me strength of brain and body at this age. Thanks be to Him." What would he do, Tosun continued, if he were to become president of Turkey? He would seek "help of money and seed for some of our farmers." Where would he want to live if he could not live in Turkey? "Nowhere. I was born here, grew old here, and hope God will permit me to die here."

The village grocer, "the only unfarming person and the only merchant in the village," presented Tosun with a different picture of his contentment: "I have told you I want better things. I would have liked to have a bigger grocery shop in the city, have a nice house there, dress nice civilian clothes. . . . I am born a grocer and probably die that way. I have not the possibility in myself to get the things I want. They only bother me."

What would he do as president of Turkey? "I would make roads for the villagers to come to towns to see the world and would not let them stay in their holes all their life." If he could not live in Turkey, where would he live? "America, because I have heard that it is a nice country and with possibilities to be rich even for the simplest persons."

We might see the chief and the grocer as illustrating the before and the after of the industrial or technological revolution. The chief's "What could be asked more?" his limited view of what he would do as president of Turkey, and his lack of interest in living outside Turkey suggest his limited level of aspirations. By contrast, the grocer's "I have told you I want better things," his desire to make roads for the villagers if he were president of Turkey, and his interest in living in America all suggest that he has been caught up in the "revolution of rising expectations" accompanying the industrial revolution. We might note that those rising expectations include both material wants ("a bigger grocery shop in the city") and nonmaterial desires ("I would make roads for the villagers to come to towns to see the world"). As a result, we have for the grocer but not for the chief a large gap between his aspirations and the fulfillment of those aspirations. And the grocer views his failed aspirations negatively ("They only bother me").

Figure I-1 below, "The Invisible Crisis" (Phillips and Johnston, 2007: 15)—is a schematic diagram suggested by the example of the grocer and the chief. It portrays two changes accompanying the shift over five centuries from preindustrial to contemporary society. There is the accelerating upper curve depicting a "revolution of rising expectations," or rapidly increasing material and nonmaterial aspirations. And there is the slowly increasing lower curve depicting the fulfillment of those expectations or aspirations, to some extent as a result of the industrial revolution's achievements. Looking at both curves at the same time, we can note the result: an accelerating gap between aspirations and their fulfillment. We might see the chief as being located at the left side of Figure I-1, where the aspirations-fulfillment gap is small. And we might see the grocer's location as not far from the right side, where that gap is large. Such a diagram does not present new evidence, but it can clarify our understanding of this contrast between the grocer and the chief by integrating a vast number of changes.

It is indeed tempting to see this example from the Lerner cross-national study along with Figure I-1 as some kind of explanation of the forces producing an accelerating aspirations-fulfillment gap. For have we not located those forces within the long-term scientific and industrial or technological revolutions? And isn't it obvious that most of us moderns are grocers of Balgat in many ways because of our aspiration-fulfillment gap and our being bothered by not being able to close that gap? This is no more than a hypothesis, however, accompanied by an illustration. And although it is supported by Durkheim's analysis, we have yet to dig very deeply into the actual complexity of what has been going on for five centuries in relation to aspirations and their fulfillment. The range of phenomena that we are discussing is huge, since it involves nothing less than the full extent of human material and nonmaterial aspirations as they have been developing over five hundred years. Because of the complexities involved within this kind of hypothesis, it is almost unique among hypotheses within the social sciences, since they focus on a much narrower range of phenomena. Nevertheless, it is most useful to put forward such a general hypothesis, granting the great difficulties involved in testing it. It helps us to "see the forest for the trees." By so doing, it yields insight into the nature of our worldview or fundamental assumptions about the world, including ourselves.

The difficulty of understanding the increasing aspirations-fulfillment gap is magnified by its largely invisible nature, granting that it is still more visible than a worldview. It is one thing, for example, to focus on the causes of World War II, with all of the visible phenomena associated with that war. But it is quite another thing to focus on the causes of an aspirations-fulfillment gap, especially one that has been developing over centuries. Despite such difficulties, however, it remains essential that we make progress in understanding the forces producing this gap. As I have already suggested, this aspirations-fulfillment gap is tied to both contemporary society's fundamental assumptions as well as its basic problems, such as: (1) our failure to provide the education needed to understand modern problems that is

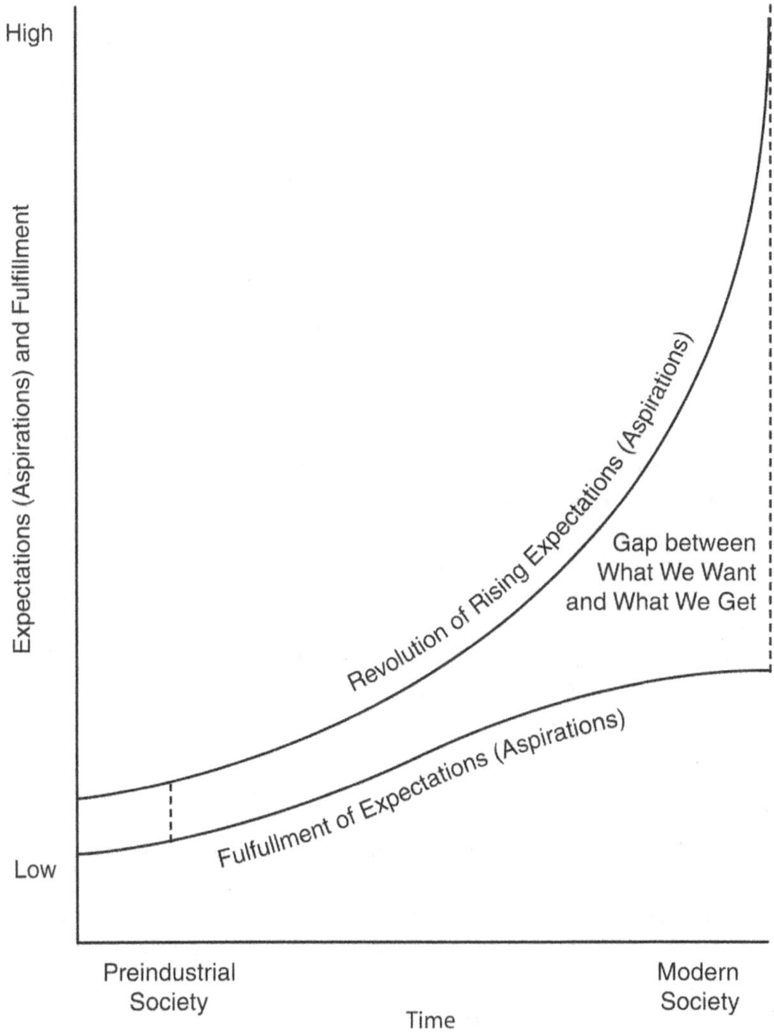

Figure I-1. The Invisible Crisis: The Escalating Gap Between Expectations and Fulfillment.

required by democratic ideals; (2) the U.S. ideal of establishing democracies in the Middle East along with resulting failures; (3) the environmental ideal of reversing global warming accompanied with limited progress; (4) the ideal of achieving greater worldwide economic equality along with the increasing gap between the rich and the poor; (5) the history of "unanticipated consequences" that have yielded the reverse of initial aspirations, such as the impact of the Treaty of Versailles; and (6) the ideals of social scientists for understanding human behavior coupled with their actual achievements.

Theoretical Framework

Given the enormous complexity of the threatening problems throughout contemporary society, as discussed, it is essential to develop a systematic framework for learning about their nature. Figure I-2, "The Escalating Invisible Crisis of Contemporary Society," suggests key relationships among the four major phenomena that have been examined up to this point, granting the actual existence of much greater complexity. Figure I-2 is meant only to begin to integrate much of the vast amount of knowledge available throughout the social sciences. Whether such integration continues will depend on ongoing efforts by the full range of specialists throughout these fields. All of the concepts discussed are quite general and largely invisible to contemporary social scientists. Even "personal and social problems" is largely invisible, since social science research centers on particular problems rather than on problems in general. I hope to make all of these phenomena more visible as I proceed throughout this book. If indeed they are important forces that are threatening contemporary society, then it is essential that we understand their nature if we hope to confront our problems effectively. If social scientists and others remain unaware of these forces, however, then I see a continuation of our increasing social problems up to the point of no return.

Following a broad approach to the scientific method, it is useful to employ three major aspects of language in examining each of the four phenomena in Figure I-2: dichotomy, gradation, and figures of speech. Although these were discussed initially in the preface—given their importance—further discussion is in order. As for dichotomy—referring to either-or thinking—this is what all languages emphasize. Every word, such as "word," divides the universe into two categories: whatever the word denotes (e.g., all the words in the universe), on the one hand, and everything else in the universe, on the other hand. Dichotomously, then, I might contrast the broad approach to the scientific method outlined in this book with a narrow approach. I might also contrast an evolutionary worldview with a stratified worldview, a small aspirations-fulfillment gap with a large one, and solving versus failing to solve personal and social problems. Gradationally, we might also see all of these phenomena as matters of degree, where we are located at some point on a line connecting the two poles of a dichotomy. For example, one individual's or one group's aspirations-fulfillment gap may be somewhat greater than that of another individual or group. As for figures of speech—which directly invoke such senses as sight and hearing—they may be illustrated by Edna St. Vincent Millay's poem about "a meteoric shower of facts," with no "loom" to "weave" those facts into "fabric." Millay uses these figures to describe our traditional scientific method, and as a result of her direct references to our sensory experiences of sight, she is able to communicate her ideas in a highly effective way.

By learning to use all three of these aspects of language—whether in scientific research or in everyday life—we open up to the incredible potential that our complex language offers to us humans, distinguishing us from

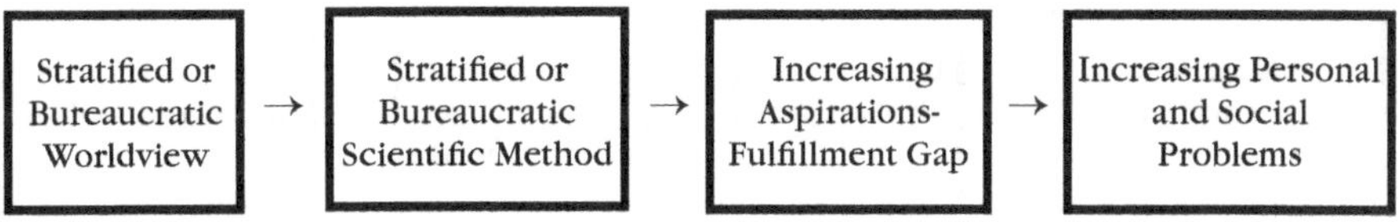

Figure I-2. The Escalating Invisible Crisis of Contemporary Society

all other organisms in the known universe. Unfortunately, however, we make use of our linguistic capacities to a very slight degree. For example, figures of speech are most important for effective communication, given their orientation to our sensory experiences. Yet they are largely excluded from the biophysical and social sciences because of their relatively inexact nature. Biophysicists generally look down on statements that cannot make use of the gradational potential that mathematics offers. And social scientists generally emphasize dichotomies, just as all of us do in our everyday thought and speech, by contrast with a gradational orientation. It is this limited development of our ability to make use of our linguistic capacities that helps to explain the origins and limitations of our stratified or bureaucratic worldview and our traditional approach to the scientific method.

In contrast with such a limited approach to language, by using all three linguistic capacities we are able to make visible what would normally remain invisible. For example, Figure I-2 emphasizes a gradational approach to language with its implication that small changes on the left side will yield similar changes on the right side. It is most useful for all of us to add this way of thinking to our ordinary emphasis on dichotomy. Yet at the same time, we can see the importance of dichotomies linked to Figure I-2, just as I've emphasized in the previously mentioned written materials: a broad versus a narrow approach to the scientific method, an evolutionary versus a stratified worldview, a small versus a large aspirations-fulfillment gap, and solving versus failing to solve personal and social problems. Of course, we can also add figures of speech, such as Millay's use of figures like the lack of a "loom" to "weave" our "meteoric shower of facts" into "fabric," which characterize a traditional approach to the scientific method. As a result of using this broad approach to language in Figure I-2, we learn to see that our stratified or bureaucratic worldview is working to yield our stratified or bureaucratic scientific method. We also learn to see the role of that scientific method working to yield our widening aspirations-fulfillment gap. Also following Figure I-2, we learn to see the result: increasing personal and social problems.

As a result of making visible crucial forces yielding our increasing personal and social problems—as depicted in Figure I-2—we can begin to understand how to change this situation. Figure I-3, "Confronting the Escalating Invisible Crisis of Contemporary Society," for example, gives us a vision of how to reverse the situation depicted by Figure I-2. We can make good use of the most powerful problem-solving device that we humans have

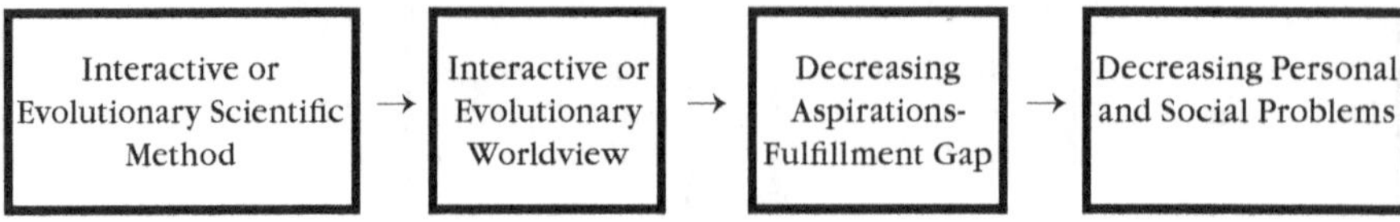

Figure I-3. Confronting the Escalating Invisible Crisis of Contemporary Society

invented aside from language itself: the scientific method. Only it would have to be a scientific method that invokes the full potential of our scientific ideals, by contrast with our present stratified or bureaucratic approach to that method. By so doing, we can—gradationally, one step at a time—chip away at our present worldview and move it in an interactive or evolutionary direction. That movement in turn should yield a narrowing of our aspirations-fulfillment gap and decreasing personal and social problems, following the findings of *The Invisible Crisis of Contemporary Society.*

Following the pendulum metaphor discussed in the preface, we can begin to understand how Figure I-3 can continue to work over time so as to yield an ever-smaller aspirations-fulfillment gap as well as a continued decrease in personal and social problems. For that metaphor suggests that an interactive or evolutionary scientific method will yield ever greater successes in solving problems, and those successes in turn will yield ever more commitment to employing that method as a basis for solving problems. Similarly, the pendulum metaphor yields insight into how Figure I-2 yields ever-increasing problems. Given a stratified or bureaucratic worldview and approach to the scientific method, their failures to solve problems—well illustrated by what is generally happening at this time in history—will yield ever greater commitment to the only tools that are presently visible, namely, that stratified or bureaucratic worldview and scientific method. And that in turn will work to yield ever more problems.

Yet we should be aware that human behavior is incredibly complex, and that Figures I-2 and I-3—along with Figure I-1—are greatly oversimplified, granting that they do succeed in conveying an overall perspective. For example, the scientific method can skip the intermediate steps depicted in Figures I-2 and I-3 and affect problems directly. Further, each of the concepts in these three figures is by no means a unitary phenomenon. Thus, for example, an approach to the scientific method—as well as each of the other concepts in these figures—may actually be made up of a set of forces that work together in complex ways, granting that its overall direction might be bureaucratic or, by contrast, evolutionary. Given this situation, research into such complexities would prove to be most useful for helping us to move from Figure I-2 to I-3 and, as a result, to narrow the aspirations-gap depicted in Figure I-1.

This book focuses not just on the scientific method and the full potential of language as used by professionals but also on their use in everyday life, whether by professionals or by the rest of us. What is involved is an approach with both short-term tactics and a long-term strategy. In the short term, any momentary experience can help us to make a choice between two alternative approaches to the scientific method, worldviews, aspirations-fulfillment gaps, and abilities to solve problems. To take seriously what appear to be trivial experiences, it is essential that we lower our aspirations in the short run so that we no longer overemphasize dichotomy (e.g., completely changing our approach to the scientific method or our worldview). But we must also learn to remain satisfied with the gradational orientation implied by "one small step." Voltaire wrote in 1772 that "the best is the enemy of the good," implying that a focus on dichotomy can easily get in the way of a short-term strategy that emphasizes one small improvement after another. We might note here the importance of the slogan, "One day at a time" for Alcoholics Anonymous groups. By trying to change something all at once rather than with one step at a time, we can easily fall flat on our faces. "The best" is what we might want to achieve ultimately, such as changing our worldview, but that emphasis can be the enemy of making some progress in that direction—"the good"—if it motivates us to take too large a step at a given time. And it is also important to make use of figures of speech, like those used by Edna St. Vincent Millay, to teach us the significance in broad terms of our taking one small step or one day at a time.

This combination of short-term tactics and long-term strategy in our approach to the problems of everyday life is carried forward throughout this book. For example, although readers will generally be Westerners—especially given my use of the English language—I bring forward Eastern thought in almost all of these chapters. And in the final chapter I develop the idea of an "East-West strategy" for confronting the problems of everyday life. That strategy centers on lowering one's aspirations in the short run so as to bring them closer to fulfillment, following Eastern thought. Once the gap between aspirations and their fulfillment is narrowed, the individual can learn to continue to raise both, following Western thought. It is this East-West strategy that is further elaborated in *Bureaucratic Culture and Escalating Problems.*

The concluding chapter of this book—chapter 11—also brings forward another approach to the problems of everyday life—"deep dialogue"—an orientation also elaborated in *Confronting Fundamental Social Problems.* All of us have conversations with others throughout our lives, yet how many of those conversations yield important learning by the participants, learning that works to develop the capacity of the individual for solving problems? "Deep dialogue" points in this direction. The chapter makes use of the ideas of the educator Paulo Freire, who emphasized the importance of dialogue. The idea of "deep dialogue" is an effort to build on his ideas.

Plan of This Book

Given the general framework in this book—which helps to systematize what I've written in this introduction as well as in previous books—how do we proceed more specifically in chapters 1 through 11? In each of the eleven chapters, I illustrate both short-term tactics and a long-term strategy, culminating in my discussion of an "East-West strategy" in chapter 11. Although each chapter emphasizes a different area of behavior—for example, social structures, the individual's personality structures, the momentary situation—a wide range of phenomena is necessarily involved within each chapter. My focus throughout is on nothing less than showing how this strategy and these tactics can actually help individuals and societies move from a stratified or bureaucratic to an interactive or evolutionary worldview, cultural paradigm, or metaphysical stance. Following Figure I-3, this requires using the scientific method in everyday life and should work to narrow the aspirations-fulfillment gap and address fundamental, increasing, and threatening problems throughout contemporary society.

Yet if we are indeed to learn how to use "the scientific method in everyday life"—and to pay attention to the other phenomena in Figures I-2 and I-3—then we must open up to the full range of our day-to-day experiences. It is here that *The Invisible Crisis of Contemporary Society: Reconstructing Sociology's Fundamental Assumptions* can help us. For those fundamental assumptions made by the sociologist are much the same as the basic assumptions that all of us make in modern society. They range over all of the structures—or phenomena persisting over time—in the universe: physical, biological, social, and personality. We may see those structures not as static entities but as products of history and change, just as Figure I-1 emphasizes the changes from preindustrial to contemporary society. Assumptions also include the incredible complexity of any given momentary situation. More specifically, chapter 1 centers on physical structures and chapter 2 on biological structures. Chapters 3, 4, and 5 focus on personality structures; 6, 7, and 8 on social structures; and 9, 10, and 11 on the situation.

More specifically, the use of three chapters each rather than only one chapter for personality structures, for social structures, and for the situation—rather than for physical and biological structures—is essential to get at the complexity of human behavior. Biophysical science has taught us a great deal about physical and biophysical structures. Combined with technological fields like engineering and medicine, biophysical science has transformed our world over the past five centuries. Yet at the same time our failure to develop social science along with social technologies has yielded a world that is becoming more and more threatening to the future of the human race. Unless we learn to shift our priorities from biophysical to social science, we have little hope of understanding those threats, much less confronting them effectively. The focus of this book on personality and social structures, as well as on the situations that we humans all experience in everyday life, illustrates that change in priorities. We might note here that social technologies need not be limited to the work of political figures, educators, social workers, and other specialists. We are all technologists or problem solvers in our everyday lives. What we problem solvers require, however, is ever-increasing understanding of the complexities of human behavior along with personal and social problems. And that is exactly what a broad approach to the scientific method coupled with a focus on social science offers us.

To understand more fully the nature of social science and social technology in our everyday lives, we might look to some figures of speech. Lyman Frank Baum's novel *The Wizard of Oz* (Baum, 1960) was the basis of that unforgettable 1939 film, where Judy Garland played Dorothy, Frank Morgan was the Wizard, Jack Haley played the Tin Woodsman without a heart, Ray Bolger was the Scarecrow whose head was filled with straw, and Bert Lahr portrayed the Cowardly Lion, who might be seen as lacking the hand that is required to act decisively in the face of danger. We might see, then, "heart," "head," and "hand" as three fundamental components of all human behavior, including science and technology. "Heart" suggests a focus on a scientific problem; "head," a focus on scientific concepts or ideas for understanding the problem; and "hand," the scientific and technological solution of the problem. The three chapters on personality structures emphasize, respectively, "heart," "head," and "hand." The same is true for the three chapters on social structures and the three chapters on the situation. But it is crucial that this is no more than a matter of emphasis. Just as our own heart, head, and hand are all part of a single organism and thus must work together, so must we take all three into account in any one of these chapters.

We can invoke science and technology in the organization of not only these chapters but also within each chapter. Each chapter is divided into two sections, with the first emphasizing a problem and the second emphasizing a direction for solving that problem, following the nature of the scientific method. The first major heading will be a component of our stratified worldview, and the second one, a component of an evolutionary worldview. For example, chapter 3 with its focus on "heart" within personality structures is divided into "Alienation" and "Emotional Expressiveness." Chapter 4 with its focus on "head" within personality structures is divided into "Stratified Worldview" and "Evolutionary Worldview." Chapter 5 with its focus on "hand" is divided into "Addiction" and "Meaningful and Expressive Rituals."

A key component of a broad approach to the scientific method is reflexivity, just as Mills and Gouldner emphasized the importance for the would-be scientist to look inward as well as outward. Language itself is our most powerful instrument for reflexivity, if only we learn to become aware of our own role in using language whenever we think, speak, or write. This orientation should work to flesh out the ideas in the various chapters. For example, writing this book requires me to confront my own past experiences. Over the years I've written three editions of a textbook—plus another textbook—on research methods in the social sciences (1966, 1971, 1976, 1985). Those books emphasized a combination of both short-term tactics and long-term strategy, and they required me to confront the nature of the scientific method. I've also written two introductions to sociology (1969, 1979), pushing me to learn a broad approach to human behavior. Overall, I've been exposed to both the shocking limitations of social science as it is presently practiced and the incredible potential of social science to help all of us at this time in history. Yet only in the twenty-first century am I beginning to succeed in integrating these experiences and putting on paper the results of my efforts. The result is my sense of the vast and threatening problems facing contemporary society as well as a direction for solving those problems.

It is most important—given my emphasis on the metaphors of "head," "heart," and "hand" and on the individual's use of an "East-West strategy" and "deep dialogue"—that this emphasis on personality structures should not be taken to be a dismissal of the importance of social structures. Given that I am attempting to reach a wide audience, it is essential that I emphasize concrete examples that have to do with the behavior of the individual. Such examples too often are lacking throughout the literature of sociology, which tends to emphasize the importance of social structures and de-emphasize that of personality structures. Yet both are most important for an understanding of human behavior. My own emphasis on the importance of both can be seen in the table of contents. By contrast, it is a one-sided focus on either personality structures or social structures that illustrates a stratified or bureaucratic worldview.

Which will it be, Armageddon or the continuing evolution of the individual's abilities to understand, to express emotions, and to solve problems? Will we choose to do no more than complain about those problems? Will we continue to obsess about choices such as between Coca-Cola and Sprite, as did—presumably—most of the passengers on United 93? Will we continue to bury world and personal problems, all the while playing our games of trivial pursuit? Will we avoid the barrage of news from the media about the increasing threats to the human situation? Or will we become increasingly aware of those threats along with the dangerous game of avoidance that we are playing? Will we proceed to use that awareness to seek out an understanding of the forces that are threatening us at this time in history as well as ways we can develop to confront those forces effectively? Will we become motivated to confront those forces, such as by following the "Let's roll!" conviction of some of those passengers on United 93, based on their awareness of the gap between their wants and the fulfillment of those wants? And will we also act, taking one step at a time, to confront ever more effectively the full range of fundamental problems that threaten us?

Given the incredible complexity of human behavior, is it reasonable to believe that we human beings can actually take that complexity into account in our day-to-day behavior? If sociologists presently are very far from integrating their knowledge—let alone social scientists in general—does it make sense to believe that all the knowledge of human behavior can indeed be integrated? Can we expect politicians, leaders, and problem solvers in all walks of life, along with everyone else, to abandon their own deep-seated convictions about how to solve personal and world problems in favor of social science knowledge? Is there really any hope of confronting effectively massive, escalating, and life-threatening problems throughout the world, given our past record of failures?

There are, of course, no guarantees in science or in life. Yet we human beings have come a very long way if one takes into account the process of biological evolution over billions of years coupled with the achievements of *Homo sapiens.* Unfortunately, we have developed our tools of language and the scientific method in a one-sided way. Yet we can continue to learn how to develop those tools without any limit. Our language sets us apart from all other phenomena in the known universe. And the scientific method, with all of its incredible potential, has as yet barely been applied to human behavior. We humans, as a result, appear to have infinite capabilities. We can learn to use these tools to confront our problems ever more effectively. Our destiny is in our own hands at this time in history. We can choose between Armageddon and evolution.

Granting that we all make that choice, it is the choices made by social scientists that I believe are of crucial importance at this time in history. For it is we far more than any other group who are responsible for developing understanding of the complex forces that are now threatening the future of the human race. Will we choose to continue to fiddle while Rome burns, avoiding any responsibility for learning the nature of those forces and communicating that understanding to the powers that be and everyone else? Or will we wake up to the increasing threats to humanity in the modern world and follow the lead of those passengers on United 93 in an effort to confront those threats effectively?

Part II

Physical and Biological Structures

Social scientists generally pay little attention to physical and biological structures, given their specialized interest in social and personality structures. Within sociology, for example, the fields of human ecology and demography do indeed center on physical and biological structures, respectively. Yet sociologists in these fields generally remain isolated from their colleagues in other fields of the discipline. As we shall see, it is crucial for us to take into account physical and biological phenomena to understand human behavior in general and human problems in particular. More specifically, we shall come to see that our understanding of what causes our large and increasing gap between aspirations and their fulfillment requires us to learn about the nature and impact of physical and biological structures. Whatever else we humans are, we are made up of physical structures that are also biological structures. Granting that our capacities extend far beyond those of all other forms of life, we also share a great deal with other organisms. And we also share characteristics with our physical universe. To ignore our physical and biological structures is also to pay little attention to our long evolutionary past, a past that has worked to shape us into what we are today. If we wish to understand ourselves we must learn about that past no less than our present situation.

Chapter 1, "Flatland Versus Spaceland/Timeland," centers on the physical structures of space and time. The chapter title is based on a science-fiction story about very limited two-dimensional creatures living in Flatland. We shall contrast that limited orientation to those physical dimensions—characteristic of a stratified worldview or metaphysical stance (Flatland)—with the broad orientation of an evolutionary worldview (Spaceland/Timeland). This contrast illustrates our sizeable gap between aspirations and their fulfillment. On the one hand, we like to think of ourselves, by contrast with other organisms, as conscious of space far beyond our present location and conscious of the distant past along with the far future, corresponding to residence in Spaceland and Timeland. Yet are we quite similar to other organisms in that we focus on our momentary scene almost all the time, like the Flatlanders, yielding a gap between what we would like to be and what we in fact are? If so, how can we learn to move from Flatland to Spaceland and Timeland? To the extent that we can accomplish this, we will also be able to move from a stratified to an evolutionary worldview.

Chapter 2, "Invisible Man Versus Visible Man," emphasizes biological rather than physical structures. *Invisible Man* is Ralph Ellison's 1947 novel about a young black man who was a victim of prejudice and discrimination: "I am invisible, understand, simply because people refuse to see me." Yet to what extent are we all invisible to one another and even to ourselves? By contrast, the idea of "visible man," is oriented to actually perceiving the complexity and uniqueness of every single individual. The key question that chapter 2 poses is, how do we move away from prejudice and discrimination, away from an oversimplified and stereotyped view of others and ourselves, and learn to take into account that complexity and uniqueness? We know from social science studies something of the widespread nature of such patterns as racism, sexism, classism, ageism, and ethnocentrism. Those patterns illustrate the treatment of people not as individuals but rather as members of a negative category of individuals. Chapter 2 suggests a direction away from a stratified worldview with prejudice and discrimination to an evolutionary worldview that opens up to the uniqueness and possibilities of every individual.

It is a broad approach to the scientific method and to language itself that can enable us to build on figures of speech like "Flatland," "Spaceland," "Timeland," "Invisible Man," and "Visible Man." From Figure I-2 we emphasize the potential impact of the scientific method on the crucial yet relatively invisible phenomena of "worldview," "aspiration-fulfillment gap," and "personal and social problems." From Figure I-1 we sketch the nature of that "aspirations-fulfillment gap." More specifically, the introduction has helped us contrast the broad approach to the scientific method suggested by C. Wright Mills (1959) in *The Sociological Imagination* with the narrow and traditional approach emphasized throughout the social sciences. It is that broad approach that we saw as pointing toward an evolutionary worldview, a narrowing of the aspirations-fulfillment gap and a reduction of personal and social problems. And it is that same approach that we can see in chapters 1 and 2 as pointing away from Flatland and Invisible Man and toward Spaceland/Timeland as well as visible man.

CHAPTER 1

Flatland Versus Spaceland/ Timeland

MY APPROACH WITHIN THIS CHAPTER AND ALL OF THE CHAPTERS TO FOLLOW parallels that of the scientific method. The first section, "Flatland," defines the problem at hand: the spatial and temporal limitations of the human being. The second section, "Spaceland/Timeland," points a direction for solving that problem. That direction makes use of the broad approach to the scientific method outlined in the introduction. It is an approach emphasizing a focus on a problem, corresponding to our focus in this chapter. It also involves moving up language's ladder of abstraction to general or abstract concepts like "worldview" and "aspirations-fulfillment gap." Equally, it involves moving down that ladder of abstraction to more specific concepts that invoke the ordinary experiences that we all have in everyday life. Those ordinary experiences include what we sense—such as what we see, hear, feel, taste, and smell—as suggested by figures of speech like "Flatland," "Spaceland," and "Timeland." This approach to the scientific method also involves a reflexive approach, where the investigator looks internally at self as well as at external phenomena. To illustrate this latter element of a scientific method broad enough to apply to the complexities of human behavior, I am writing this book partly to learn more about how to shift my own worldview further toward an evolutionary metaphysical stance.

Flatland

William Butler Yeats's poem "The Second Coming" was written in 1921, not long after the end of World War I, with its millions lying dead on the battlefields of Europe. That war was a severe blow to the Enlightenment optimism of the eighteenth and nineteenth centuries, and Yeats's poem from his "Second Coming," published in 1921 reflects the decline of that optimism in the twentieth century, as illustrated by these lines:

> Turning and turning in the widening gyre
> The falcon cannot hear the falconer;
> Things fall apart; the centre cannot hold;
> Mere anarchy is loosed upon the world.

We might note the breadth of Yeats's orientation to the physical dimensions of time and space. He is writing about developments that have been taking place over centuries, and also about developments that encompass contemporary societies in general.

By contrast with Yeats's breadth in time and space, we have people's generally limited orientation to time and space. To put this contrast into sharper relief, we turn to Edwin A. Abbott's *Flatland: A Romance of Many Dimensions* (1884/1952), an early science fiction novel written by the headmaster of an English school. Abbott describes a two-dimensional world populated by Triangles, Squares, Circles, and other figures from plane geometry who move about on their plane but remain unable to rise above it or sink below it. A three-dimensional Sphere visits Flatland and carries off a Square above his plane and into a three-dimensional world.

The Square becomes enchanted by his new vision of the world, and when the Sphere returns him to Flatland he is most eager to tell his countrymen of the wonders of a three-dimensional world. Yet he remains unable to convince anyone of the existence of Spaceland. Worse, as a result of his seditious remarks he is imprisoned for life, and he voices this lament:

> Hence I am absolutely destitute of converts, and, for aught that I can see, the millennial Revelation has been made to me for nothing. Prometheus up in Spaceland was bound for bringing down fire for mortals, but I—poor Flatland Prometheus—lie here in prison for bringing down nothing to my countrymen. Yet I exist in the hope that these memoirs, in some manner, I know not how, may find their way to the minds of humanity in Some Dimension, and may stir up a race of rebels who shall refuse to be confined to limited Dimensionality. (Abbott, 1884/1952: 102)

It is indeed tempting for us to look down on the creatures of Flatland who imprisoned the Square, yet are we not similar to them? They rebel at anyone who challenges their fundamental assumptions or worldview, yet isn't that our own pattern of behavior as well? For example, I'm claiming here that we are in fact potentially creatures of Timeland, a four-dimensional world, yet we generally behave as if we exist only in the present moment. For example, how frequently do we think about our own past experiences or the history of society, let alone apply what we've learned in those situations to our present experiences? Other organisms without our highly developed language are necessarily doomed to focus on the present moment, but are we so different from those organisms with our own focus on the present? And even if we think of the past and the future to some extent, just how far back do we go in personal or world history, and how far forward do we go? Is there a huge gap between our aspirations to live in Timeland and the fact that we generally fail to do so?

If we focus on world history, let us note the remark of the philosopher George Santayana: "Those who cannot remember the past are condemned to repeat it." Just how much do we remember of past history, and just how much have we learned from that past? What have we learned from the decline and fall of the Roman Empire? From the Crusades? From the American and French revolutions? From World War I, which was billed as the "war to end all wars"? From World War II? From the Holocaust? From the Vietnam War? From 9/11? From the clashes in Israel and Palestine? From the current wars in Iraq and Afghanistan? The continuation of wars throughout human history suggests that we have learned very little about how to achieve peace.

Up to this point I have emphasized the physical dimension of time. We might do well now to focus on space. To illustrate, the psychologist Robert Sommer in *Tight Spaces: Hard Architecture and How to Humanize It* (1974) describes an experiment he performed on the Berkeley campus of the University of California. Sommer's research assistants arrived at classrooms just after classes were leaving and just before students entered for the next classes. They converted the straight rows of chairs into circular arrangements and then proceeded to observe the results:

> The data here were not what we expected. We had hoped to compare participation in the circular arrangement with participation in sessions when the chairs were in straight rows. In this case, we found that twenty of the twenty-five classes rearranged the chairs back into straight rows before the class began! This was true for small classes as it was for larger ones, and for social science classes as well as natural science classes. What was most discouraging was that students were often the instigators in returning to row arrangements, sometimes before the instructor even arrived. In several cases the instructor made negative remarks and the students practically jumped to rearrange the room according to his wishes. (Sommer, 1974: 87–88)

Why were these results "discouraging" for Sommer? A circular versus a straight-row arrangement encourages far more interaction among the students, and it points away from an emphasis on persisting hierarchy or social stratification with the teacher on top and the students on the bottom. Thus, a circular arrangement is closer to fulfilling educational ideals based on research about how students learn, and it is also closer to democratic cultural values. How are we to explain these results, assuming that students and instructors—certainly on the Berkeley campus—have learned these values? Why would 20 out of 25 classes rearrange the chairs so as to point away from democratic values? Another "depressing" finding for Sommer was that instructors monopolized 40 or more minutes in all of the classes studied, leaving students with very little time for discussion, once again pointing away from democratic ideals. As before, we see a large gap between ideals and actual behavior.

Let us probe more fully into this abandonment of ideals. Arthur J. Vidich and Joseph Bensman in the late 1950s studied "Springdale," a town in upper New York State, and published their results as *Small Town in Mass Society: Class, Power and Religion in a Rural Community* (1960). Springdale township had a population of only 3,000 at the time that Vidich and Bensman did their study. It served largely as a farm trading center, and it was quite dependent on outside forces such as corporate and governmental officials far from Springdale. One focus of the study was on how Springdalers managed to cope with fundamental gaps between their aspirations and the degree to which those aspirations were actually fulfilled:

> *The technique of particularization* is one of the most pervasive ways of avoiding reality. It operates to make possible not only the failure to recognize dependence but also the avoidance of the realities of social class and inequalities. The Springdaler is able to maintain his equalitarian ideology because he avoids generalizing about class differences. The attributes of class are seen only in terms of the particular behavior of particular persons....
>
> The realization of lack of fulfillment of aspiration and ambition might pose an unsolvable personal problem if *the falsification of memory* did not occur, and if the hopes and ambitions of a past decade or two remained salient in the present perspective. But the individual, as he passes through time, does not live in spans of decades or years. Rather, he lives in terms of seasons, days and hours and the focus of his attention is turned to immediate pressures, pleasures and events.... As they [hopes and aspirations] are in process of disappearing, other thoughts of a more concrete and specific nature occupy the individual's attention, and new goals are unconsciously substituted for those that are being abandoned.... As a consequence, his present self, instead of entertaining the youthful dream of a 500-acre farm, entertains the plan to buy a home freezer by the fall. (1960: 299, 303; italics added)

It appears that "the technique of particularization" works to narrow one's exposure to the physical dimension of space and that "the falsification of memory" works to narrow the individual's exposure to the physical dimension of time. Both of these are procedures for burying one's awareness of problems that one would rather not face or is unable to solve. Sigmund Freud might use the term "repression" for what the Springdalers did. Yet the results of this behavior are by no means completely negative, if we accept Vidich and Bensman's conclusion:

> Because they do not recognize their defeat, they are not defeated. The compromises, the self-deception and the self-avoidance are mechanisms that work; for, in operating on the basis of contradictory, illogical and conflicting assumptions, they are able to cope in their day-to-day lives with their immediate problems in a way that permits some degree of satisfaction, recognition and achievement. (1960: 320)

There is, however, a better way, one that does not require us to give up on obtaining "some degree of satisfaction, recognition and achievement." Thomas Hardy, the English novelist and poet, wrote these lines in 1902 in "De Profundis," a poem whose Latin title means "from the depths": "If way to the Better there be, it exacts a full look at the worst." Hardy's title catches a central idea of the scientific method: we must first become aware of a problem before we can begin to solve it. Here I am also invoking Figure I-2 where a broad approach to the scientific method can alter our worldview. This focus on the scientific method is the basis for the organization of this chapter and the ones to follow: we begin with a section on a problem, and we conclude with a section pointing toward procedures for solving the problem. The approach is much like that of a swinging pendulum—as suggested in the preface—where a swing in one direction points toward understanding a problem. And a swing in the other direction points toward a solution. To gain needed momentum for arriving at a solution, we must first swing the pendulum as far as possible toward understanding the nature of the problem. Alternatively, if we have no idea as to how we might solve a problem, then we have little momentum for facing up to the nature of the problem. The alternative we generally adopt in that case is simply avoidance of the problem, as illustrated by the Springdalers.

Yet there is an alternative to such increasing problems. The deeper we move into understanding a problem, the more momentum we have for making progress on solving it. And the closer we are to solving the problem, the more momentum we have for understanding it more fully. These possibilities are sketched in Figure I-3, where the aspirations-fulfillment gap is narrowed and there are decreasing problems, with both occurrences based on a change from a stratified or bureaucratic scientific method and worldview to an evolutionary scientific method and worldview. Now that we have gained at least some momentum from understanding the problems of our Flatland behavior, let us proceed at this point toward a vision of Spaceland and Timeland.

Spaceland/Timeland

> Everyman, look to the harvest! It is the layman's responsibility to be aware of his own aspirations and those of the group to which he belongs. It is for him to choose the vision he will follow and to take responsibility for carrying it out.... No man or woman is exempt from taking up the challenge. social scientist, intellectual, artist, leader, middleman of any breed, and the Common Man (and Woman) to whom, after all, this century belongs—each must ask himself, what is my vision of the future? And what am I doing about it? ... Man has the capacity to dream finer dreams than he has ever succeeded in dreaming. He has the capacity to build a finer society than he has ever succeeded in building.... Here lies the real challenge! There are among us even now dreamers and builders ready to repeat the age-old process of splitting the atom of time, to release the Western world from its too-long imprisonment in the present. Then man will once again be free to "seek the city which is to come."
>
> —*Fred Polak,* The Image of the Future, *1973: 305*

Fred Polak was a Dutch sociologist who surveyed the "images of the future" in the major religions within the history of Western society. This excerpt comes at the end of his book. He concluded that "the more powerful the image of the future is, the more powerfully it acts in determining the actual future." Polak saw images of the future as varying in their ability to actually shape the future. But exactly what is the nature of those images of the future that have proved to be most powerful historically, that is, most successful in shaping the future? Lawrence Busch, a contemporary sociologist, examined major social movements throughout Western history in his doctoral dissertation (1974) and summarized his conclusions in a later article:

1. An image of the future must be holistic if it is to achieve widespread acceptance....
2. A successful image of the future must provide the promise of the resolution of the anomalies and contradictions of the existing order....
3. The future must be constructed in the present, not the future....
4. A successful image of the future must provide an escape from the existing order, but it must find that escape within the existing order....
5. A successful image of the future must provide an operationalizable methodology for the individual....
6. All successful images of the future are structured....
7. A meaningful image of the future must involve the mundane.... (1976: 29–36)

Based on his analyses of many social movements, Busch gives us a key requirement for achieving fundamental changes in society, much like changing a worldview or narrowing the aspirations-fulfillment gap: developing a "successful" image of the future. As for (1), the image must be holistic or broad: "It cannot restrict itself to a particular problem but must forthrightly address *all* the problems plaguing the present." (2) clarifies (1): it must promise not only to address that range of problems but must promise to solve them. (3) specifies the necessity of action in the present to resolve those problems, by contrast with specifying that actions take place at some future time. (4) clarifies (3), ruling out any retreat to some spatially remote area such as the utopian communities of the nineteenth century or the retreat communes of the 1970s in the United States. (5) specifies the importance of a methodology that any individual can use to fulfill the image of the future, corresponding to my own idea of developing a scientific method that any individual can use to solve problems in everyday life. (6) cites the importance of structure, as illustrated by the Constitution of America's founding fathers, or any written document providing a set of rules that can be applied over and over again in one situation after another. As for (7), that clarifies the nature of the methodology specified in (5): it must be sufficiently concrete to apply to any everyday situation whatsoever.

Busch's analysis of the components of an image of the future that can be most influential in actually changing the future is nothing less than a general description of the components of a scientific method broad enough to yield understanding of human behavior and human problems, as discussed in the introduction. It is, then, a prescription for a scientific method that all of us—including those who have received Nobel prizes in science along with all of our most eminent biophysical and social scientists—can learn to use in one momentary situation after another to narrow our aspirations-fulfillment gaps. It is also a prescription for changing our worldview or metaphysical stance from one that is stratified to one that is evolutionary. If we look to Figure I-3, it is a focus on the incredible potential of a broad approach to the scientific method that is located in everyone's hands, not just among professional scientists.

If indeed this broad approach to the scientific method will prove to be a passport to the future, taking us from Flatland to Spaceland and Timeland, then let us proceed to examine its nature in greater detail. The components of the Web and Part/Whole Approach to the scientific method were discussed initially in *Beyond Sociology's Tower of Babel: Reconstructing the Scientific Method* (Phillips, 2001) and later in *The Invisible Crisis of Contemporary Society* (Phillips and Johnston, 2007):

1. Defining a fundamental problem (as all problems are linked, no problem is trivial),
2. Moving up language's ladder of abstraction to very general concepts,
3. Moving down language's ladder of abstraction to very concrete concepts,
4. Integrating knowledge, and

5. Performing a reflexive analysis within an evolutionary or interactive worldview.

Busch's points (5) and (6) outline a methodology that is "structured"; that is, it can be repeated, implying that it is recorded and is systematic, as illustrated by the outline of the Web and Part/Whole Approach to the scientific method. This is, of course, only a general outline, with details provided not only in the two books just cited but also in two edited volumes (Phillips, Kincaid, and Scheff, eds., 2002; and Phillips, ed., 2007). It is a methodology that is not fixed in concrete but has been developing on the basis of experience. Busch's point (2) with its focus on addressing fundamental problems is much like (1) in the Web Approach: we must begin by defining a fundamental problem. As for Busch's point (1) with its idea that an image of the future must be "holistic," that is achieved by (2) along with (4) and (5) in the Web Approach by moving very far up language's ladder of abstraction, even to concepts like "stratified worldview" and "evolutionary worldview" that get at our metaphysical stance. Those very general concepts enable us to encompass the range of phenomena that Busch requires, making it possible to integrate knowledge. Busch's point (7), oriented toward "mundane" phenomena, is much like the Web's point (3), with its focus on moving very far down language's ladder of abstraction. Finally, Busch's points (3), (4), and (5), which call for an image that individuals can use immediately to solve problems and proceed to change society, point up the importance of using the scientific method not just to gain understanding but also to solve problems in society. In other words, Busch suggests that successful images of the future call for social technologies based on the scientific method, just as engineers and physicians presently base their efforts on physical and biological science.

In our effort to move from Flatland to Spaceland, we should understand that we live in an interactive universe where it is not possible to completely separate any part from the rest of the universe, such as attempting to put something in a "perfect vacuum." There is no such thing as a perfect vacuum, for infinitesimal particles of matter from outside the vacuum's container will collide with the molecules that make up the container's walls, thus transmitting energy to the contents of the container. Similarly, the matter inside the container will interact with those walls and transmit energy outside the container. Of course, some phenomena, such as organisms by contrast with inanimate matter, are more interactive than other phenomena. For example, a tree requires a continuing supply of sun, water, and carbon dioxide for it to continue living, and it sends out a continuing supply of oxygen. We humans, given our language, are by far the most interactive creatures in the known universe, for we can think about and be influenced by events that occurred billions of years ago or that could possibly occur in the distant future. And we can also be influenced by what we read about the experiences of others, whether living or dead. Yet to the extent that we remain victims of our stratified worldview—thus emphasizing the situations we are in from one moment to the next—we do not live up to our human capacities. By dipping into Spaceland only sometimes, the structure of our spatial perspective keeps us in Flatland. Thus, we fail to fulfill capacities we have developed on the basis of billions of years of biological evolution.

We humans are the product of a world that has been evolving for fourteen billion years. Every time we look at ourselves we might well take into account that huge stretch of time. Our very existence embodies those fourteen billion years, and not only the four billion or so years of biological evolution. Given our own importance within the universal scheme of things, we would do well to become most "reflexive" about the importance of our own personal history. For that personal history has shaped our personality structures along with the social structures that presently are influencing us.

A key idea introduced in the introduction and expanded in this chapter, the potential of language, is basic to understanding the nature of the scientific method. My own introduction to the importance of language within the scientific method occurred during my first years at Columbia College, when I was a premedical student and took courses in chemistry, physics, and mathematics. It was then that I learned the importance of general or abstract concepts like "force," "nucleus," and "valence." Mathematics was also important, but it was no more than a tool to connect phenomena in a highly systematic way. Here were efforts to deal with problems involving the physical dimensions of time and space, and it was general concepts that saved the day. I never lost that insight. And when I changed my major to sociology during my final years—as a result of taking courses with C. Wright Mills—I was appalled at how little attention social scientists paid to the importance of a systematic approach to the general concepts developed by social scientists. Instead, many social scientists came to see mathematics as central to the scientific method. Yet mathematics is best employed on problems much simpler than those having to do with human behavior. Mills himself was not guilty of this failure to understand how the scientific method had actually worked in the physical sciences. For example, in a study of textbooks on social problems while still a graduate student at the University of Wisconsin, he drew this conclusion: "The level of [linguistic] abstraction which characterizes these texts is so low that they seem to be empirically confused for lack of abstraction to knit them together. They display bodies of meagerly connected facts, ranging from rape in rural districts to public housing, and intellectually sanction this low level of abstraction" (1943: 166).

Neither was Alvin W. Gouldner—whose orientation to a reflexive approach was quoted in my introduction—guilty of neglecting the importance of the general concepts developed by social scientists. After the publication of *The Coming Crisis of Western Sociology* (1970) where he introduced ideas about reflexivity, he reacted to a critical review of that book several years later. I repeat here several lines from a quote in the preface:

> At decisive points the ordinary language and conventional understandings fail and must be transcended. It is essentially the task of the social sciences ... to create new and "extraordinary" languages, to help men learn to speak them, and to mediate between the deficient understandings of ordinary language and the different and liberating perspectives of the extraordinary languages of social theory. (1972: 16)

Have social scientists in fact followed Gouldner's advice of 35 years ago that they should "create new and 'extraordinary' languages," and "mediate between the deficient understandings of ordinary language and the different and liberating perspectives of the extraordinary languages of social theory"? And on the basis of this "language-reform and language-creating business," have they moved toward "creating a new culture"? In *The Coming Crisis of Western Sociology,* Gouldner also indicated his interest in the importance of "background assumptions," "world hypotheses," and "metaphysics," implying that a new culture would involve absolutely fundamental changes, such as a changed worldview. There is little question but that those changes have not occurred. To illustrate, presently there are no fewer than 46 distinct Sections of the ASA with relatively little communication among the members of different Sections, let alone communication with social scientists in other disciplines. Introductory textbooks throughout the social sciences present hundreds of concepts, but almost invariably fail to distinguish central from less important concepts. And just as we move from one chapter to the next, earlier concepts are almost completely ignored as we proceed to move from one specialized area of research to another, where never the twain shall meet. The result is a lack of any framework of organized concepts, such as our understanding of the relationship between "flower" and "roses," "daffodils" and "hyacinths." Rather, there are disconnected concepts. There is no integrated "extraordinary language" that we can use in our day-to-day lives that links to our everyday language and helps us to understand the world and solve problems.

As for "creating a new culture," only a tiny minority of social scientists attempt to become what is called "public intellectuals"—like Mills—in an effort to reach a wide audience. Rather, generally they are content to communicate only within their own very narrow specialized area, considering it a major accomplishment if they ever manage to communicate to other specialists within their discipline. They like to congratulate one another if someone ever manages to place evidence of the existence of social scientists in, say, *The New York Times.* But this is light years away from what Mills achieved, let alone communicating regularly and powerfully to the public at large. As for developing a vision of what Gouldner called "a new culture" and filling out that vision with specific structures within our range of institutions, social scientists have presented to one another critiques of a great deal of contemporary culture along with specific institutions. But those criticisms generally are highly specialized and often contradict one another, failing to yield integrated knowledge such as we find within the physical and biological sciences. Neither are those criticisms generally followed up with constructive ideas in any detail about how those problems might be solved. As a result, changes toward a broader approach to the scientific method that is used in everyday life have remained no more than a dream.

A portion of the "extraordinary language" that Gouldner called for is to be found in the part and chapter headings throughout this book. Those concepts are by no means the only ones that might constitute such a language. But they do illustrate an extraordinary language in their systematic relationships, their abstract nature, their derivation from the literatures of the social sciences (by contrast with being restricted to ordinary language), and the fact that they can work to integrate a very wide range of knowledge throughout the social sciences and beyond. This is by no means an argument for abstract or general concepts at the expense of concrete or specific ones. On the contrary, the Web and Part/Whole Approach to the scientific method calls for us to shuttle up and down between levels of linguistic abstraction. The initial emphasis of this research approach, as communicated in *Beyond Sociology's Tower of Babel,* was on abstract concepts. It was described as the Web Approach. However, Thomas J. Scheff had developed and illustrated what he called a "part/whole" approach to the scientific method in three monographs that he published in the 1990s (1990, 1994, 1997). That approach emphasized the importance of getting deeply into the momentary situation with the aid of concrete sociological concepts, just as novelists generally accomplish this with the aid of everyday concepts. Yet despite his focus on the momentary scene, he did not abandon a holistic approach to society as a whole. His influence yielded a change from the Web Approach to the Web and Part/Whole Approach to the scientific method.

The previous paragraphs have to do with social science language and, to a lesser extent, with the language of the biophysical sciences. But what about the language of the humanities, such as what is communicated in novels, poetry, film, and plays? Shouldn't such insights be brought forward in the following chapters? For example, hasn't Shakespeare contributed a great deal to our understanding of human behavior and human problems? Shouldn't we give far more than lip service to such contributions? Readers might note my quoting of Edna St. Vincent Millay—along with usage of Lyman Frank Baum's *The Wizard of Oz*—in the introduction, and also my use of Yeats's "The Second Coming" and my reliance on Edwin Abbott's *Flatland* in this chapter. Thus, I've already used literature for illustrative purposes. Let us recall as well my analysis of the dichotomous, gradational, and figurative potentials of language at the end of the section "Theoretical Framework" in the introduction. For it is all three that we can learn to fulfill to a much greater extent, thus helping us to develop a broad or evolutionary approach to the scientific method as well as an evolutionary worldview. At this point, let us proceed to expand on that idea, given its enormous importance.

Let us turn to Laurence Perrine, a well-known author of textbooks on poetry and other literature:

> But it is not primarily to communicate information that novels and short stories and plays and poems are written. These exist to bring us a sense and a perception of life, to widen and sharpen our contacts with existence. Their concern is with *experience.* The poet . . . creates significant new experiences—significant because focused and formed—in which the reader can participate and that he may use to give him a greater awareness and understanding of his world. Literature, in other words, can be used as a gear for stepping up the intensity and increasing the range of our experience and as a glass for clarifying it. (1973: 4)

It appears that literature accomplishes much the same thing as good social science, giving the reader "greater awareness and understanding of his world." Language, whether used by social scientists or poets, is a most powerful tool for solving problems. We can see this clearly in the focus of the scientific method on beginning with awareness and definition of a problem and then proceeding to make progress toward solving that problem. But poetry and literature accomplish something else that prose generally does not accomplish, or at least achieves to a lesser degree. Perrine gives us a clue when he claims that "novels and short stories and plays and poems . . . exist to bring us a sense and a perception of life," where the words "sense" and "perception" refer directly to sensory experiences like seeing, hearing, touching, tasting, and smelling. From this perspective, literature takes us very far down language's ladder of abstraction to such very specific phenomena. Yet Perrine is claiming that literature also moves us very far up that ladder, since it gives the individual understanding of "his world," and this understanding would include the individual's worldview. Another aspect of literature, following Perrine, is its ability to increase the "intensity" of our experience, thus emphasizing the "heart" or emotional experience no less than the "head."

Perrine continues with this illustration from *Macbeth*:

> To-morrow, and to-morrow, and to-morrow
> Creeps in this petty pace from day to day
> To the last syllable of recorded time;
> And all our yesterdays have light fools
> The way to dusty death. Out, out, brief candle!
> Life's but a walking shadow, a poor player,
> That struts and frets his hour upon the stage
> And then is heard no more. It is a tale
> Told by an idiot, full of sound and fury,
> Signifying nothing.
>
> —*Shakespeare,* Macbeth, *1606, Act 5, Scene 5, Line 16*

Shakespeare's passage appeals to our senses at a low level of abstraction. For example, he gives us images for the nature of life: a "brief candle," "a walking shadow," "a poor player." In addition to appealing to our sense of sight, he also appeals to our sense of hearing, in common with poetry in general. We have his alliterations or repetitions of sounds when the words are spoken and not just read: "To-morrow, and to-morrow, and to-morrow," "petty pace," "day to day," "dusty death," and "poor player." There is also the rhythm of alternating unaccented followed by accented syllables in each line, similar to our experiencing musical repetition, the beat of our hearts, the pulse of our blood, or our breathing in and out. But Shakespeare also moves up to a very high level of abstraction, for his focus on the nature of life itself has to do with our fundamental assumptions, worldview, or metaphysical stance. He achieves this, following Perrine's analysis, with "sudden leaps," illustrated by "Life's but a walking shadow, a poor player," instead of "climbing up [a stairway] one step at a time'" He is not required to present evidence, slowly and laboriously, as he moves up the ladder of abstraction from the concrete "walking shadow" to the abstract concept of "life," as is the scientist. But by "sudden leaps" he helps us to become aware of problems, which constitute the first step of the scientific method. Is life in fact much like a brief candle? A walking shadow? A poor player? Each question can be pursued much further with other questions, given the concrete image that is involved. With respect to "a brief candle," for example, can we see life as a phenomenon that can easily be snuffed out, that does not last long, that begins and ends in darkness, that flames in intensity, that gives off light and energy, that is active and colorful, and that gets shorter and shorter?

I presented a view of poetry—and literature in general—of moving up and down language's ladder of abstraction in *Beyond Sociology's Tower of Babel* (2001: 22). There I depicted the languages of physical and biological science, of social science, and of literature as similar in enabling the reader to shuttle up and down that ladder. I also wrote about their differences. Physical science (and biological science to a lesser extent) emphasizes language's gradational component—that is, seeing phenomena as matters of degree rather than in an either-or manner—with its frequent usage of mathematics. Social science, with its usage of both ordinary language and what Gouldner has called "extraordinary language," emphasizes linguistic dichotomy, for every word we use (like "candle") divides the world in two (candles and noncandles). Finally, literature emphasizes figures of speech. As a result, I suggested that the scientist should make full use of figures of speech along with dichotomous and gradational thought to communicate effectively to self and others.

Figures of speech generally are able to reach our awareness more easily than other language because they dip into nothing less than our patterns of perception from one moment to the next. We are all organisms first of all and users of language only second of all. And it is our perception that is involved here. It is exactly these patterns of perception that are the focus of chapter 2, "Invisible Man Versus Visible Man." We are biological creatures who see and hear before we learn language. Of course, language works to change the way we see, hear, touch, taste, and smell. But just because of that shaping by language, we cannot afford to ignore the overriding importance of our sense experiences from one moment to the next. Unfortunately, our stratified worldview has taught us to separate an understanding of human biology, including our sense experiences, from an understanding of human behavior, yielding the competition we now see between explanations based on biological and social science. We lamely try to pull the two together with the old saw that we are a product of both heredity and environment, but that does not go very far in helping us to understand human behavior. Crucial questions are neither asked nor answered. For example, does our worldview affect our perception? Does a stratified worldview work to prevent us from becoming reflexive and paying attention to our own sensory experiences? Can an interactive or evolutionary worldview work to change our very patterns of perception? Can our patterns of perception in turn work to alter the nature of our worldview? And can they help us to become aware of problems from one moment to the next, like our residence in Flatland rather than Spaceland and Timeland? Our biological nature includes much more than our patterns of perception, and we shall at least touch on other aspects of human biology. But our focus will remain on perception, given its overriding importance in helping us to learn about and change our behavior.

CHAPTER 2

Invisible Man Versus Visible Man

WE MOVE IN THIS CHAPTER FROM A FOCUS ON THE PHYSICAL UNIVERSE to an exploration of the living world, centering in particular on human perception. This includes not just how we see the world but also our senses of hearing, touch, taste, and smell. Yet, although our attention turns to biological structures, physical structures remain crucial to an understanding of the complexity of human behavior. But why center on perception? To address human perception is by no means to suggest that human biology trumps the impact of our experiences in society. Rather, it is to make some room for biological structures that do in fact influence our behavior. Yet those structures are not fixed for all time. They can be changed to some extent. For example, I suggested at the close of chapter 1 that our worldview might influence the way we perceive the world. If this is true, then perhaps our patterns of perception can work to influence the nature of our worldview. That is the basis for my interest in focusing on perception: we may be able to change the way that we sense the world so as to encourage a shift from a stratified to an evolutionary or interactive worldview.

Near the end of chapter 1 I began to give perception its due in analyzing the language of literature and, in particular, figures of speech. Those figures are basic to the enormous impact that poetry, drama, and literature in general can have on us, for they point us directly to our sense experiences. Indeed, figures of speech have infiltrated language in general, just as we say, not only "I understand," but also "I see." The images suggested by "Flatland," "Spaceland," and "Timeland," for example, should help us to understand the contrast between a stratified and an evolutionary or interactive worldview. Yet a focus on perception can do more for our understanding of that contrast as well as of how to shift from one to the other. I noted in the introduction the importance of structures, as illustrated by the physical, biological, personality, and social structures that we center on throughout this book. Structures are crucial to understanding human behavior and human problems because they continue to influence us from one moment to the next, just as a chair or the moon doesn't suddenly disintegrate, just as a leopard doesn't suddenly change its spots, and just as we humans don't suddenly change our worldview, our patterns of hierarchy, or the nature of our personalities. Another structure is our patterns of perception: we perceive phenomena from one moment to the next to the next. By contrast, our thoughts are a sometime thing. And if our patterns of perception encourage a stratified worldview, perhaps we can alter those patterns so that they encourage an evolutionary or interactive worldview.

In common with all of the other chapters, this chapter is divided into a section emphasizing a problem ("Invisible Man") and a section oriented to a solution ("Visible Man"). Just as I used literary materials like *Flatland* to clarify and emphasize ideas in chapter 1, the same is the case for this chapter. Ralph Ellison's novel, *Invisible Man* (1947), suggests not only the existence of racial prejudice and the existence of other patterns of social stratification like sexism, ageism, religious prejudice, classism, and ethnocentrism. *Invisible Man* also suggests the relative invisibility of all human beings, as influenced by our stratified worldview or metaphysical stance. By contrast, the quotation from Walt Whitman's "Song of Myself" (1892/1983) after the section heading "Visible Man," suggests quite a different view of the individual. Instead of failing to see the unique nature of the individual, Whitman "celebrates" himself, referring to his breathing, his sense of smell, the sights and sounds he experiences, and his sense of touch. And he does not neglect the human being's linguistic capacities, for he sees all of us with the potential to learn "the origin of all poems."

Invisible Man

> I am an invisible man. No, I am not a spook like those who haunted Edgar Allan Poe; nor am I one of your Hollywood-movie ectoplasms. I am a man of substance, of flesh and bone, fiber and liquids—and I might even be said to possess a mind. I am invisible, understand, simply because people refuse to see me. Like the bodiless heads you see sometimes in circus sideshows, it is as though I have been surrounded by mirrors of hard, distorting glass. When they approach me they see only my surroundings, themselves, or figments of their imagination—indeed, everything and anything except me.
>
> Nor is my invisibility exactly a matter of a biochemical accident to my epidermis. That invisibility to which I refer occurs because of a peculiar disposition of the eyes of those with whom I come in contact. A mat-

> ter of the construction of their inner eyes, those eyes with which they look through their physical eyes upon reality ... you're constantly being bumped against by those of poor vision ... out of resentment, you begin to bump people back. And, let me confess, you feel that way most of the time. (Ellison, 1947: 3–4)

Ellison's hero, a young African American, is perfectly visible as a physical object and as an organism. But he is invisible as a unique human being with a history of experiences different from those of any other human being. He is seen as a black man and, as a result, he is treated stereotypically and in a hierarchical way as a member of the Negro race. Yet that categorization drastically conflicts with American cultural values emphasizing the fundamental equality of all human beings. From the Declaration of Independence we have this statement: "We hold these Truths to be self-evident, that all Men are created equal.... " From Lincoln's Gettysburg address we have these words: "... we here highly resolve that these dead shall not have died in vain—that this nation, under God, shall have a new birth of freedom, and that government of the people, by the people, and for the people, shall not perish from the earth." From the Old Testament of the Bible, in the book of Genesis, we have this: "And God said, Let us make man in our image, after our likeness.... " Not white men or black men, but "man." There is, then, a large gap between egalitarian ideals or values and the negative ways in which Ellison's hero is treated. This is yet another illustration of a large aspirations-fulfillment gap. Yet by contrast with the Springdalers, discussed in chapter 1, Ellison's hero does not react with a "falsification of memory" of his ideals. Instead, he reacts to his frustration: "out of resentment, you begin to bump people back. And, let me confess, you feel that way most of the time."

This idea of the invisibility of human beings was carried much further to include the invisibility of all phenomena by Plato in the fourth century B.C.:

> imagine men to be living in an underground cave-like dwelling place, which has a way up to the light along its whole width, but the entrance is a long way up. The men have been there from childhood, with their neck and legs in fetters, so that they remain in the same place and can only see ahead of them, as their bonds prevent them from turning their heads. Light is provided by a fire burning some way behind and above them. Between the fire and the prisoners ... there is a path across the cave and along this a low wall has been built, like the screen at a puppet show in front of the performers who show their puppets above it.... See then also men carrying along the wall, so that they overtop it, all kinds of artifacts, statues of men, reproductions of other animals....
>
> Do you think, in the first place, that such men could see anything of themselves and each other except the shadows which the fire casts upon

> the wall of the cave in front of them? . . . Altogether then, I said, such men would believe the truth to be nothing else than the shadows of the artifacts? . . . the men below had praise and honours from each other, and prizes for the men who saw most clearly the shadows that passed before them, and who could best remember which usually came earlier and which later.
>
> —*Plato's* Republic, *1974: 168–169*

This allegory of prisoners in a cave in the *Republic* is perhaps the best-known image from the entire history of Western philosophy. Plato is suggesting that we are all much like those prisoners in the cave who can see nothing more than shadows on the cave's wall. We cannot even see ourselves or the other prisoners. All of the phenomena that we see are far from reality. Yet can Plato's idea be correct? How could we possibly be seeing shadows and not be seeing what is really there? We all trust our vision, for otherwise we would not risk such everyday behavior as crossing a busy street, driving a car, walking down a stairway, or doing most of the things that we do. There is the twentieth-century proverb, "One picture is worth ten thousand words." If we cannot trust our senses, then what can we trust?

Yet perhaps our senses are not completely deficient but only limited in what they report to us. Ellison's hero was indeed a black man, and those who saw him were correct in that respect. But when it comes to seeing human beings, and not merely physical objects like cars or stairways, our vision communicates an incomplete and even distorted story. But what about the physical objects that we see? Does our vision betray us there, as Plato suggests? Perhaps Plato does indeed make sense. Certainly we see the cars and the stairways quite accurately. But do we at the same time fail to see our own bodies? Is our sight of objects from one moment to the next accompanied by our own invisibility? Let us recall from chapter 1 the figures of speech of Flatland, Spaceland, and Timeland. To what degree does our vision work to narrow us spatially, instead of helping us move into Spaceland? To what degree does that vision work against our moving from Flatland to Timeland, where we would see the momentary situation from the perspective of months, years, centuries, or even millions and billions of years? We might recall here the technique of particularization and the falsification of memory used by the Springdalers to limit their exposure to space and time.

Since Ellison's time, and certainly since Plato's time, our problem of seeing or sensing only a portion of what is actually there has increased dramatically, just as the aspirations-fulfillment gap in Figure I-1 has accelerated with the change from preindustrial to contemporary society. At this time in history we are all besieged by advertising wherever we go: on television and radio for 24 hours a day; on cars, buses, planes, and trains; in newspapers and magazines; on billboards in the countryside; and even on people's clothing and skin. John Berger examined the impact of advertising some 35 years ago in his *Ways of Seeing* as well as in a BBC television series:

> The spectator-buyer is meant to envy herself as she will become if she buys the product. She is meant to imagine herself transformed by the product into an object of envy for others, an envy which will then justify her loving herself ... the publicity image steals her love of herself as she is, and offers it back to her for the price of the product.
>
> Publicity speaks in the future tense and yet the achievement of this future is endlessly deferred. How then does publicity remain credible—or credible enough to exert the influence it does? It remains credible because the truthfulness of publicity is judged, not by the real fulfillment of its promises, but by the relevance of its fantasies to those of the spectator-buyer. Its essential application is not to reality but to daydreams.
>
> Publicity turns consumption into a substitute for democracy. The choice of what one eats (or wears or drives) takes the place of significant political choice. Publicity helps to mask and compensate for all that is undemocratic within society. And it also masks what is happening in the rest of the world. (Berger, 1972: 134, 146, 149)

Whatever else publicity or advertising does, it works to narrow us in time and space, keeping us in Flatland rather than Spaceland and Timeland, and working much like the Springdalers's technique of particularization and falsification of memory. We become the victims of hearing or seeing one message after another, taking our attention away from our own lives and turning it to the desires of the advertiser. And let us not forget the power of seeing and hearing. We have seen some of that power in the analysis of figures of speech in chapter 1. Figures are able to move very far both down and up language's ladder of abstraction. Advertising can communicate directly to our senses of vision and hearing, and some advertising also manages to reach our other senses as well. As for moving far up language's ladder of abstraction, advertising can and does promise the world and generally uses no ponderous research to back up its messages. If only we buy this or that product, we can achieve sexual gratification and intimate relationships, respect from others, wealth, power, fame. Indeed, we can have any fulfillment whatsoever that the mind of the advertiser can conceive of. And if we do have such unfulfilled desires—as illustrated by Figure I-1—then all of us are ripe for the picking.

Following Berger's argument, advertising does not merely tap into the aspirations-fulfillment gap that we already have, it works to increase that gap by creating dissatisfaction with what we have and who we are: "Publicity speaks in the future tense and yet the achievement of this future is endlessly deferred." We become simultaneously dissatisfied with what we are at present and envious of what we might become if we buy the product advertised. Yet we never achieve the fulfillment that advertising promises. For one thing, those promises generally are either grossly exaggerated or simply false. Further, even if a given promise is actually fulfilled, we are besieged with ever more advertising that works to widen rather than narrow our aspirations-fulfillment gap.

Berger also argues that advertising "turns consumption into a substitute for democracy." Our choice among dozens of available skin lotions, shampoos, or toothpastes comes to be a substitute for the fundamental political choices that we should be making in the face of mammoth problems within contemporary society. Just as advertising itself distracts us from those problems and our passivity in relation to them, so do our trivial choices of products carry further that distraction and passivity. Advertising becomes, then, another technique of particularization and another process for falsifying memory, as if we ourselves were not already deeply involved in those procedures. Advertisers make it easier for us to engage in such distraction and passivity by promising us the world in exchange for our purchases, but those promises remain hollow. Our choice among toothpastes, just as in the case of the Springdalers's choices, "permits some degree of satisfaction, recognition and achievement." Yet at the same time we are learning, ostrich-like, to bury our heads in the sands while waiting for catastrophe.

Yet another problem with advertising is the extension of its techniques to the political arena. If there is no searchlight from the social sciences to train on the advertisers of products, neither is there a searchlight to train on the advertisers of candidates for public office. Here we are confronted by yet another huge aspirations-fulfillment gap: between our democratic cultural values requiring that the people will make informed choices when they vote, and the severe limitation of our understanding of what choices will in fact prove to be useful. That limited understanding is a product not only of the limitations of our formal educational institutions, but also of every single one of our other institutions, for they all teach a stratified or bureaucratic worldview or metaphysical stance. And that stance is itself tied to a very limited approach to the scientific method that favors narrow specialization with very limited communication among those specialized and subspecialized areas of knowledge.

To illustrate the sale of political candidates along with political issues, we turn to Ian Buruma's review of Frank Rich's *The Greatest Story Ever Sold: The Decline and Fall of Truth From 9/11 to Katrina* (2006a):

> As a former theater critic, Frank Rich has the perfect credentials for writing an account of the Bush administration, which has done so much to blur the lines between politics and show business. Not that this is a unique phenomenon; think of Silvio Berlusconi, the media mogul and master of political fictions, or Ronald Reagan, who often appeared to be genuinely confused about the difference between real life and the movies.... What is fascinating about the era of George W. Bush, however, is that the spinmeisters, fake news reporters, photo-op creators, disinformation experts, intelligence manipulators, fictional heroes and public relations men posing as commentators operate in a world where virtual reality has already threatened to eclipse empirical investigation.... This kind of official arrogance is not new, of course, although it is perhaps

> more common in dictatorships than in democracies. What is disturbing is the way it matches so much else going on in the world: postmodern debunking of objective truth, bloggers and talk radio blowhards driving the media, news organizations being taken over by entertainment corporations and the profusion of ever more sophisticated means to doctor reality. (Buruma, 2006: 9–10)

It appears, with "spinmeisters, fake news reporters, photo-op creators, disinformation experts, intelligence manipulators, fictional heroes and public relations men," coupled with "postmodern debunking of objective truth, bloggers and talk radio blowhards driving the media, news organizations being taken over by entertainment corporations and the profusion of ever more sophisticated means to doctor reality," that we are rapidly moving toward the world that Plato depicted in his allegory of prisoners in a cave. If all of this is indeed happening more and more, then what we are coming to see is nothing more than shadows on the cave wall. It even brings to mind the lies told by dictators throughout history to preserve their power. And it calls up short the idea of our democracy as pointing toward "government of the people, by the people, for the people." How are we to have meaningful elections if candidates are sold in this way and if issues are distorted to win political victories? What trust can we have in our free press if entertainment becomes more important than news, and if reporters are becoming eager to conform to the wishes of the powers that be? How can an understanding of contemporary problems be communicated widely if the airwaves have been taken over by those who are advertising candidates and their own narrow spin on issues? How are those postmodernists who have claimed that the scientific method itself should not be favored over other efforts to arrive at truth to be answered? What is to be done about the flooding of the airwaves with manipulated information? *The Greatest Story Ever Sold* is thus about the creation of a false reality. There is an abstract reason for this creation of a false reality that can be added to those concrete ones: the invisibility of the social sciences, and the failure of social scientists to integrate their knowledge so as to give reporters something far more than the ability to quote this or that "expert" on a given problem. If Rich blames the press primarily, I put the primary blame on the social science community. How could such a false reality have been created? It has been created largely because social scientists have not done their job of providing reporters and everyone else with a solid foundation for understanding what has been happening, based on the wealth of knowledge hidden in our libraries coupled with the application of that knowledge to the current scene. More specifically, if social scientists have not been using a broad approach to the scientific method that enabled them to accomplish all of this, and if their stratified worldview got in the way of both using such a method and communicating the results of their investigations as widely as possible, it becomes unfair to place the primary blame on reporters.

The creation of a false reality is yet another example of a huge gap between aspirations and their fulfillment. This time, the gap is between our interest in truth, which is, of course, vital for solving any and all of our problems, and the false reality that is communicated to us. One possible solution to this gap along with all of our other aspirations-fulfillment gaps takes us back over 2,500 years to the time of Gautama Siddhartha Sakyamuni, commonly known as the Buddha, or the Enlightened One (Kaplan, 1961: 237-266). He was born into the family of a ruler of one of India's smaller principalities and was raised as a prince of the royal house. After having grown up leading a sheltered life, he left the palace for the first time in his early twenties and experienced encounters with a very old man suffering from severe infirmity, an invalid suffering from his disease, a corpse about to be cremated, and a monk seeking his salvation. Gautama was so shocked by a realization that old age, sickness, and death were the common lot of the human race, an understanding that he had been sheltered from while living in a palace, that he decided to follow the path of the monk and to seek his salvation. In his "Great Renunciation" he renounced his wealth and position, left his wife and child and friends, and directed his life toward achieving enlightenment.

Gautama began by studying philosophy, but he found no answers there. He then proceeded to mortify his body almost to the point of death, but once again was disappointed with the result. After years of effort he sat down under what later came to be known as the bo tree or tree of enlightenment, determined not to rise up until he had solved the great problems suffered by the human race. Finally, at the age of 35, he achieved what he had sought for so many years. He rose up from under the bo tree and proceeded to Sarnath, on the Ganges River near Varanasi in north-central India, and preached his first sermon. In the following half-century he traveled up and down the Indian subcontinent—including a return to his family, with his son becoming one of his disciples—sharing truths he had attained and setting in motion the "Wheel of the Law."

Buddhist teaching is based on the Four Noble Truths and the Eightfold Noble Path. The first Noble Truth is the universal existence of "dukkha," which has to do with pain, sorrow, and suffering, but more precisely suggests the lack of complete fulfillment of all human pursuits. There is, then, always an aspirations-fulfillment gap. Nothing is ever wholly satisfying, and this is true even for the most fortunate human beings. This is the problem that Gautama addressed, much like the focus of the scientific method on a problem. The law of causality—like the assumption of modern science, and like the Hindu law of karma—is the second Noble Truth: everything that happens is caused. If we remove the cause of dukkha, then dukkha will disappear. The third Noble Truth is the renunciation of desire, 'thirst," or aspirations, the removal of passions that will inevitably be frustrated to some degree. The fourth Noble Truth specifies the Eightfold Noble Path as the concrete direction for actually removing dukkha: right views, right aspirations, right speech, right conduct, right livelihood, right efforts, right thoughts, and right contemplation.

Is there, then, any genuine solution to the problem of our increasing aspirations-fulfillment gap or, in Gautama's terms, the experience of dukkha for all of us? Can we actually hope, for example, to root out the prejudice and discrimination that Ellison's hero encountered in the *Invisible Man*? If our manipulated reality—or shadows on the cave wall—is backed up by powerful political and economic institutions—can we hope to shine the light of enlightenment on those distortions and thus move toward truth? The musical *Chicago* makes fun of a legal system that doesn't succeed in convicting women who have killed their husbands because the defense attorney along with the murderers succeed in "razzle-dazzling" the jury, or creating a false reality much as Rich (2006) and Buruma (2006) describe. Yet *Chicago* was the product of show business, and we can have a good time laughing at the way the defense succeeded in its razzle-dazzling efforts. But if in fact we are all being razzle-dazzled about world problems and present efforts to solve them, then—as President Bush the elder declared—we are in "deep doo-doo." Is it indeed possible for us to extricate ourselves? Realistically, can we somehow succeed in learning to confront our escalating problems effectively?

Visible Man

> I celebrate myself, and sing myself,
> And what I assume you shall assume,
> For every atom belonging to me as good belongs to you. . . .
> My respiration and inspiration, the beating of my heart,
> the passing of blood and air through my lungs,
> The sniff of green leaves and dry leaves, and of the shore
> and dark-color'd sea-rocks, and of hay in the barn,
> The sound of the belch'd words of my voice
> loos'd to the eddies of the wind,
> A few light kisses, a few embraces, a reaching around of arms,
> The play of shine and shade on the trees as the supple boughs
> wag. . . .
> Have you felt so proud to get at the meaning of poems?
> Stop this day and night with me
> and you shall possess the origin of all poems. . . .
>
> —*Whitman,* "Song of Myself," *1892/1983: 23-24*

In one word, the answer to the above question is yes. It is we human beings who have created the massive problems we now face, and it is we humans who have the capacity to solve them. Let us not forget the two most powerful tools that we have invented: language and the scientific method. It is language that sets us apart from all other creatures on earth and all other structures within the known universe. And it is language that gives us the capacity to reach out in space and time, entering Spaceland and Timeland, and to interact with all phenomena within our universe. It is the scientific method that has succeeded in shaping the contemporary world, even despite our limited understanding of its nature. And it is a broader approach to the scientific method—one that enables us to build on language's dichotomous, gradational, and figurative potential—that can penetrate the mysteries of human behavior and human problems. It has taken no fewer than fourteen billion years to arrive at our present situation. Should we, at this point in time, simply give up on what appear to be our infinite possibilities as human beings for continuing to evolve? With all of the achievements of the human race, is it reasonable to declare that we cannot solve contemporary problems?

We can begin by updating Gautama's approach to the scientific method, as well as the limited understanding of that method presently guiding contemporary social science. Most important, we can alter the idea of *completely* eliminating human aspirations or "thirst" to eliminating *unrealistic* aspirations. This is illustrated by the passage from Durkheim's *Suicide* in the introduction, where he describes the greed associated with the industrial revolution: "greed is aroused without knowing where to find ultimate foothold. Nothing can calm it, since its goal is far beyond all it can attain" (1897/1951: 255). Durkheim calls not for the elimination of aspirations but rather for aspirations limited by realism. Such reasonableness is also suggested by Voltaire, also quoted in the introduction: "The best is the enemy of the good." Rather than try to fulfill aspirations that are too high, we should lower those aspirations so as to take "one small step for a man," following Neil Armstrong's statement when he stepped on the moon, and going along with the slogan for Alcoholics Anonymous groups: "One day at a time." Thus, by not abandoning the idea of the importance of reasonable aspirations, we do not violate what social science has taught us. Yet at the same time we move to reduce dukkha even though we do not violate what we've learned about emotions from contemporary social science.

We can also take other steps to update the Four Noble Truths. Gautama's sense of problem can be understood to include problems that have been emphasized in the foregoing pages, such as our limited usage of the scientific method, our stratified worldview, and our growing aspirations-fulfillment gap, as outlined in Figure I-2. His movement up language's ladder of abstraction can include the basic concepts developed throughout the social sciences, such as those listed in this book's chapter titles. His efforts to move down that ladder can open up to the detailed knowledge developed throughout the social sciences as a basis for learning how one can learn to choose "right" living in a given situation at a given point in time. Still further, the individual can point toward reflexivity, just as the Web and Part/Whole Approach calls for, so as to develop a broad approach to the scientific method. This effort to use a broad scientific method in everyday life follows the spirit of Gautama's Four Noble Truths. And it should also work to reduce dukkha, following Gautama's argument as well as my own. Those Noble Truths help us clarify our understanding of how to narrow the aspirations-fulfillment gap by developing aspirations that are realistic within the momentary situation so that dukkha is minimized. This is the idea of taking no more than one step at a time.

In the long run, by contrast, we can continue to raise our aspirations if they are fulfilled in one situation after another. With respect to Figure I-1, this idea is depicted by the curves of aspiration and fulfillment coming close together and then jointly moving upward, following the Western idea of progress or development. Yet such development need not imply the one-sided materialistic development that has yielded serious environmental problems coupled with an increasing gap between the rich and the poor, discussed in chapter 8. Rather, it can be a balanced economic development that guarantees environmental sustainability and economic justice. This may appear to be no more than a naive hope at this time in history. Yet it can become a realistic state of affairs to the degree that we gain awareness of the problems associated with our emphasis on a materialistic economy and economic injustice.

It would be difficult to think of someone other than Walt Whitman who presents such an enormous contrast with Gautama's emphasis on eliminating or reducing one's "thirst" or emotional attachments to desires. We can see in the quote from "Song of Myself" the depth of Whitman's attachments. Whitman is most eager that his readers open up to the joy of living, as he is at heart a teacher with the welfare of his readers in mind when he writes, "For every atom belonging to me as good belongs to you." Whitman's aspirations for those readers are very high, for he sees them as having the potential to "possess the origin of all poems." Yet much of this contrast vanishes once we distinguish between short-term and long-term aspirations. Our desires can reach to the skies in the long run without being unrealistic—thus following Gautama's emphasis on realism—provided that we are able to work up to those heights one step at a time by fulfilling limited desires in the short run.

The quote from "Song of Myself" helps us to understand the sensual basis of a reflexive orientation. For example, Whitman writes about "my respiration . . . the beating of my heart, the passing of blood and air through my lungs, the sniff of green leaves . . . the sound of the belch'd words of my voice . . . a few light kisses . . . the play of shine and shade on the trees . . . ." This includes sight, sound, touch, and odor. Yet he does not remain at this very low level of linguistic abstraction. He also rises very high on that ladder when he writes, "I celebrate myself" and "you shall possess the origin of all poems." For he is writing about nothing less than the continuing development or evolution of the individual. Thus, by moving far down and up language's ladder of abstraction, Whitman is using figurative language so as to enable a broad scientific method to reach more phenomena and also become more concrete, as discussed in chapter 1. That achievement of figurative language combined with a broad scientific method is crucial for responding to advertising's sale of objects and political candidates combined with its focus on a stratified worldview. Whitman, however, does more than this. He also points toward the further evolution of the individual human being.

Yet a fundamental problem with those earlier visions is that they failed to make sufficient use of the incredible potential of language and the scientific method. And that failure apparently derived from a largely invisible commitment to a stratified worldview or metaphysical stance. These are ideas suggested by Figure I-2. One step in the direction of building on that potential would be an image of the future broad enough to encompass the visions of Whitman, Polak, Mills, and Gouldner. Such a vision would also build on the earlier ideals of the Renaissance and the French Enlightenment along with the words of Thomas Jefferson's Declaration of Independence and Abraham Lincoln's Gettysburg address. I might proceed to hint at such a vision by paralleling the speech that Martin Luther King gave at the March on Washington on August 28, 1963: "I have a dream that one day on the red hills of Georgia the sons of former slaves and the sons of former slave owners will be able to sit down together at the table of brotherhood." I have a dream that:

> there will be a future for our children, our grandchildren, our great-grandchildren, and their great-grandchildren;
>
> one day we will all learn to see ourselves as children who are only just beginning to understand ourselves and our world, and we will also learn to dream about our infinite possibilities and move toward those visions one step at a time;
>
> one day we will all learn to pay close attention to the accomplishments of all peoples throughout history as well as to our own personal accomplishments, and we will also learn to pay close attention to the failures of the human race and to our own personal failures;
>
> one day we will be able to bring to the surface and reduce our stratified emotions like fear, shame, guilt, hate, envy and greed, and we will learn to express ever more our evolutionary emotions like confidence, enthusiasm, happiness, joy, love and empathy;
>
> one day we will see peace on earth and fellowship among all humans;
>
> one day we will no longer look down on any other human being; and
>
> one day we all will learn to be poets, philosophers and scientists.

Following Polak's (1973) vision and Busch's (1976) analysis in chapter 1, an image of the future must be realistic no less than optimistic if it is indeed to influence the actual shaping of the future. Let us, then, obtain a dose of realism by beginning to examine the actual nature of human biology. Although our biological nature is often viewed as placing substantial limits on our possibilities for development, the following quote from Stephen Jay Gould, an eminent biologist suggests that the reverse is true:

> We are, in a more than metaphorical sense, permanent children.... Many central features of our anatomy link us with fetal and juvenile stages of primates: small face, vaulted cranium and large brain in relation to body size, unrotated big toe, foramen magnum under the skull for correct orientation of the head in upright posture, primary distribution of hair on head, armpits and pubic areas.... In other mammals, exploration, play, and flexibility of behavior are qualities of juveniles, only rarely of adults. We retain not only the anatomical stamp of childhood, but its mental flexibility as well.... Humans are learning animals. (Gould, 1981: 333-334)

Gould sees us humans as learning animals. In the introduction we saw language and the scientific method as central to our ability to learn. Gould adds to this picture details on our biological structures, giving us the basis for learning. In particular, it is our large brain that makes possible our development and use of a complex language. Our biological evolution as vertebrates and large-brained mammals has yielded a dramatic contrast with the social insects and their specialized castes, such as workers, drones, and queens (Herrick, 1956: 172-175). Within those totalitarian social structures, the group is the cardinal unit, and the individual blindly conforms to biological impulsions. By contrast, vertebrates, mammals, and especially human beings are far more individualistic biologically. But given our stratified worldview, along with our patterns of social stratification, we have learned to imitate the social insects to a great extent. Are we humans, then, too individualistic, as many would claim? Perhaps, however, we are not sufficiently social because we have not yet developed sufficiently as individuals so as to be able to reach out, as well, to others. Perhaps we have learned to repress, like the Springdalers, not only our awareness of problems we cannot solve, but also our emotions in general, making it difficult for us to either love ourselves or love others.

Instead of writing about how biology teaches us our limitations, Gould writes about how biology teaches us our possibilities, giving us reason to welcome biological knowledge rather than to avoid it. If we look to the seven dreams that I have—which are linked closely to an evolutionary worldview—we can see how our biological nature does indeed prepare the way for us to learn to fulfill them. For one thing, it prepares the way for our complex language, which in turn opens us up to infinite possibilities for further learning and development. Biologically, our structure is like that of children, preparing us for that process of learning and development. Indeed, there is reason to believe that we can learn to evolve from one moment to the next. This process can include our learning to face up to repressed emotions like fear, shame, guilt, hate, envy, and greed. As a result, we can learn to express ourselves in constructive ways. We can also continue to dream or develop images of possible futures, and that in turn will help us to move toward those futures. Those dreams can include our learning to become poets, philosophers, and scientists. Yet nothing in our biological nature appears to require us to look down on others, to make war, or to be aggressive in our relationships. Neither does our biological structure require us to see ourselves as either more or less intelligent than others. Biologists who have emphasized how biological structures limit us may well be victims of a stratified worldview that has limited their own understanding.

It is not just biologists who have written about our limitations but a number of social scientists as well. The theories that they have developed generally have emphasized the power of the group over the individual, largely ignoring human possibilities as individuals. Generally, they have stressed the idea that those social forces will inevitability triumph over any individual efforts in other directions, with the idea of "inevitability" being most unscientific. For example, there is Karl Marx's theory of the inevitability of class warfare and the eventual triumph of workers over owners. There is also Auguste Comte—the founder of sociology—with his theory of the inevitable triumph of science over superstition. And we have the anthropologist Lewis Morgan, who saw technology over the ages—through innovations that fulfilled human needs—as, inevitably, the driving force of human evolution. Charles Darwin's fundamental breakthrough in understanding biological evolution—*The Origin of Species* (1859/1958)—has been interpreted by some as requiring, necessarily, the "survival of the fittest" and the elimination of the "unfit," thus supporting an extremely stratified orientation that contradicts humanistic cultural values emphasizing the ultimate worth of every single individual.

Yet each of these figures has contributed important ideas to our understanding of human evolution. Unfortunately, their unscientific ideas—or the misinterpretations of others as in the case of Darwin—have worked to give the very idea of human evolution a bad name, and many present-day social scientists avoid this idea. However, some contemporary social scientists have become committed to the importance of the idea of human evolution despite the history of its unscientific use or those misinterpretations of Darwin's theory, as illustrated by the formation of a new Section of the ASA, "Evolution and Sociology." If we adopt a broad scientific method, as illustrated by the Web and Part/Whole Approach, then we can come to understand and emphasize the contributions of the figures just mentioned. By focusing on their limited understanding of the scientific method, we lose out on learning from those contributions.

For example, it was Karl Marx's theories that emphasized the centrality of patterns of social stratification, illustrated by the hierarchy between owners and workers. It was Marx who wrote about the devastating effects of social stratification on the life of the worker, effects that we can generalize to all patterns of social stratification throughout society. Auguste Comte, the founder of the discipline of sociology, opened up possibilities for an extremely broad understanding of human behavior. Further, his emphasis on the overriding importance of a scientific approach to achieving that understanding has been central to the movement toward the development of the social sciences as genuine sciences. Lewis Morgan's emphasis on the importance of technology—followed by contemporary social scientists like Leslie White and Gerhard and Jean Lenski—helps us to understand just how the scientific method has proven to be such a powerful force over the past five centuries. It also helps us to understand how initial ideas that were similar to the scientific method in earlier times resulted in inventions—such as agriculture—that became influential in shaping the development of human societies. As for Darwin, a key idea in his work is the importance of very long-term interaction between organisms and their environments, for that interaction is the basis for biological evolution. Similarly, it is genuine two-way egalitarian interaction between two individuals, by contrast with a hierarchical relationship, that is also the basis for learning by those individuals.

At this point we would do well to return to the central question guiding all of these chapters, as sketched by Figure I-3: how can we learn to use a broad approach to the scientific method in our everyday lives as a basis for altering our worldview, closing our aspirations-fulfillment gap, and solving personal and social problems? Following the work of Gould and Herrick, we need not emphasize the limitations that our biological nature places on our behavior. Instead, we can emphasize the possibilities that it opens up to us. Social scientists, with the partial exception of psychologists, generally have emphasized how social structures have succeeded in shaping the individual. Here, we might think of two such powerful social structures: patterns of social stratification as well as language, which is also a social structure because it is widely shared throughout society and persists over time. There is no question but that social structures have indeed shaped individuals throughout history and that this is a most fundamental insight in understanding human behavior. Yet it is also true that individuals have shaped and can shape social structures, for all phenomena interact with one another, and human beings are the most interactive structures in the known universe.

But how might this actually be accomplished? In this chapter our focus is on perception. Perception is not merely sensation, that is, the experience of seeing, hearing, touching, tasting, or smelling. It is sensation that has an impact on how we behave in a given situation. And that impact is in turn influenced by the range of phenomena that all of these chapters address. For example, perception can work to shape how we define that situation, whether consciously or unconsciously. Following Ellison's *Invisible Man,* a white person's sensation of seeing a black man may result in his definition or labeling of that man as being less worthy, less intelligent, or less reliable than that of a white man. This example illustrates the idea that our experiences in society shape our perceptions. And those perceptions in turn shape our behavior from one moment to the next.

If we refer to the seven dreams I listed previously, dreams of our living in an evolutionary or interactive society versus our present stratified one, we can gain ideas for how we might work to change our perceptions so as to move toward an evolutionary society. Here is a listing of four dreams we might focus on:

> that one day we all will learn to be poets, philosophers, and scientists;
>
> that one day we will be able to bring to the surface and reduce our stratified emotions like fear, shame, guilt, hate, envy, and greed, and that we will learn to express ever more our evolutionary emotions like confidence, enthusiasm, happiness, joy, love, and empathy;
>
> that one day we will no longer look down on any other human being; and
>
> that one day all of us human beings will learn to see ourselves as children who are only just beginning to understand ourselves and our world, and we will also learn to dream about our infinite possibilities and move toward those visions one step at a time.

The first dream centers on our two most powerful tools: language and the scientific method. By learning to be poets, we can learn to use more figures of speech and to use them more effectively. And in so doing we can become more aware of our sensations and perceptions with more opportunity to change those perceptions. By learning to be philosophers we can learn to become aware of our worldview as well as to alter it, thus helping ourselves to move toward a worldview emphasizing a broad approach to perception. And by learning to be scientists we can develop powerful tools for solving problems, such as how to change our worldview.

As for the second and third dreams, they both have to do with our learning to confront problems we have experienced as a result of our having lived within a stratified society emphasizing a stratified metaphysical stance. With respect to the second dream, it is possible for us to learn how to perceive our own emotions, granting that this is a difficult process, especially because we've learned to repress our emotions within a stratified worldview. It is here that the first dream can aid us, helping us to focus on ourselves with the aid of figures of speech, using a scientific method that challenges an outward-oriented worldview, and learning to see ourselves as extremely important. In particular, we can learn to see our emotions, giving us a basis for shifting from stratified to evolutionary ones. In the same way, we can learn to see our stratified behavior toward others. Having a direction for a way out, that is, an evolutionary worldview, enables us to confront our problematic behavior, just as the pendulum of the scientific method swings back and forth from solving problems to confronting them more fully.

As for the fourth dream where we learn to see ourselves as children and look toward fulfilling our infinite possibilities, my doctoral dissertation found that older people who see themselves as younger are significantly more optimistic and are in better health. We don't know whether it is their self-image that affects their optimism and health, or the other way around, or some combination of the two. But this suggests that learning to see ourselves as children may in fact help us in important ways, such as giving us the optimism to change our worldview. Our broad approach to the scientific method requires that it is continuing progress on problems rather than final solutions that should be our orientation. That orientation supports the idea that we should learn to see ourselves as children and thus learn to look forward to our possibilities. We might also note that a Timeland perspective also supports such a self-image.

As we move to Part III with its focus on personality structures, we should bear in mind that perception—like other human phenomena—is a most complex process. It is affected not only by physical and biological structures but also by personality structures, social structures, and situational phenomena. We must include all of these phenomena if indeed we wish to change our worldview and, as a result, learn to make progress on the fundamental problems that threaten contemporary society. In particular, our emotions are crucial. For most of my life I've been trying to figure out just how I can learn to express my own emotions more fully. As I've learned, the problem I am facing is a profound one and extremely difficult for me or anyone else to make progress on. It is not simply that I am a Westerner, a professor, and a male, all of which point me away from emotional expression and toward emotional "alienation." It is also that my stratified worldview or metaphysical stance pushes me—like the Springdalers—to bury my emotions. If I wish to change that worldview then I must become motivated to do so, and I must retain that commitment in one situation after another. Can I along with others accomplish this? I'm convinced that this is possible. As for the procedures that we might use, we turn to chapter 3 for some answers.

Part III

Personality Structures

In a famous line from Walt Kelly's comic strip about an educated opossum who keeps up with current events, Pogo declared, after seeing litter under a tree, "We have met the enemy and he is us." This was an image that later became a poster for Earth Day. Pogo was taking off on Commodore Oliver Hazard Perry's announcement of his victory over the British in the Battle of Lake Erie on September 10, 1813: "We have met the enemy and they are ours." As we have seen in chapter 2, our stratified worldview—taught to us by our social structures—points us away from seeing ourselves. Yet an evolutionary or interactive worldview—illustrated by Pogo's orientation—requires us to look at ourselves no less than our external environment.

If, indeed, we succeed in learning to see ourselves repeatedly, we can become aware not only that we are physical structures located in time and space and that we are biological structures who perceive the world from one moment to the next, but also that we are personality structures and that we have the capacity to confront problems that lie within our very own patterns of behavior. We have learned from the social sciences about the power of social structures to shape personality structures. This is the result of what sociologists have called the process of "socialization." But within our interactive universe, and especially given the human being's enormous capacity to interact with external phenomena, the reverse occurs as well: the individual shapes social structures. The individual is shaped by the process of socialization, and in turn the individual shapes society.

But what indeed is the nature of personality structures? These are repetitive patterns of behavior of the individual. If they come to be widely shared, then they are not only personality structures but also social structures. More specifically, those patterns of behavior have to do with perception as well as—figuratively—"head," "heart," and "hand." This includes our self-image and worldview (perception), our beliefs ("head"), our values ("heart"), and our rituals ("hand"). These images, beliefs, values, and rituals all interact with one another; with physical, biological, and social structures; and with any given momentary situation, making for a great deal of complexity. Yet we can dip into that complexity with the aid of a powerful system of beliefs that goes by the label of the scientific method. By so doing we can make progress in learning to alter our worldview, self-image, beliefs, values, and rituals. And following Pogo, we can come to see our own personality structures as creating problems for us in those scenes where we are, in fact, the enemy.

We begin our focus on personality structures in chapter 3, "'Heart': Alienation Versus Expressive Orientation." It is in the heart, figuratively, that our deepest feelings or emotions lie, feelings that have worked to shape our self-image and worldview no less than our values, beliefs, and rituals. In that chapter we shall point toward raising those feelings up to the surface where we can learn their nature and change them if we choose to do so. In chapter 4, "'Head': Stratified Versus Evolutionary Worldview," we shift to a focus on our beliefs, especially our worldview combining beliefs with perceptions. Whatever is the nature of our worldview, it has helped to teach us to bury our emotions, like the Springdalers, when we are confronted by problems that we are unable to solve. Yet when those beliefs along with those emotions become visible, it becomes possible to alter them. Chapter 5, "'Hand': Addiction Versus Meaningful and Expressive Rituals," contrasts two different ways of acting in relation to our problems. Addictive actions are rituals that work against the fulfillment of our fundamental values. By contrast, we can develop meaningful and expressive rituals that accomplish the reverse.

CHAPTER 3

"Heart"

Alienation Versus Expressive Orientation

IN THIS CHAPTER THE PROBLEMS WE CONFRONT DERIVE from four major intellectual traditions: that of Sigmund Freud, Karl Marx, modern sociology—as illustrated by the work of Thomas J. Scheff and Suzanne Retzinger—as well as existentialism. We are most selective, however, with reference to each of these cases. For example, Freud wrote that "anatomy is destiny," emphasizing the idea of the overriding importance of biological structures in determining human behavior. By contrast, the neo-Freudians Karen Horney and Erich Fromm reserved an extremely important place for social structures in explaining human behavior, and it is their work that I shall emphasize. Marx centered on alienation and economic forces within the workplace, but it is Georg Simmel's much broader focus on alienation that is no less important. Modern sociology is so diverse as to have no center. Scheff provides us with a focus on emotions, a much-neglected and central phenomenon in understanding personal and social problems. And existentialism generally is communicated with a focus on despair; yet it can yield a most optimistic perspective, as illustrated by key ideas conveyed by Nietzsche and Sartre. Without such a selective approach, there would be no way to integrate, systematically, the bits and pieces of social science knowledge. By so doing, we avoid fighting wars over which theorist we should bow down to. Instead, we are able to stand on the shoulders of our earlier giants, carrying their central ideas further. This is not only the nature of ideals for the scientific method but also the way we human beings have proceeded to learn, and to develop effective social technologies or procedures for solving problems, throughout human history.

Alienation

> The second contradiction is that between the stimulation of our needs and our factual frustrations in satisfying them. For economic reasons needs are constantly being stimulated in our culture by such means as advertisements, "conspicuous consumption," the ideal of "keeping up with the Joneses." For the great majority, however, the actual fulfillment of these needs is closely restricted. The psychic consequence for the individual is a constant discrepancy between his desires and their fulfillment. (Horney, 1937: 288)

The title of the book from which this excerpt is taken, *The Neurotic Personality of Our Time,* reveals Horney's overall assessment of culture and society in the 1930s: they work to create serious problems for the individual. It is not just a small proportion of us who are affected but all of us to some degree, since we all participate in culture. The quote points to the same phenomena depicted in Figure I-1. She saw a contradiction between "the stimulation of our needs," suggesting the top curve of aspirations, and "our factual frustration in satisfying them," suggesting the bottom curve of the fulfillment of aspirations. We have only to turn to chapter 2, where we see Berger's examination of advertising, to understand that the problems Horney saw in the 1930s have been increasing over the decades: "Publicity speaks in the future tense and yet the achievement of this future is endlessly deferred" (Horney, 1937). We are besieged by ads for products, and no matter how many we might be able to afford, there are a great many others that we have not purchased. For Horney, it is this aspirations-fulfillment gap that is the basis for neurosis: learning to want things that one cannot obtain.

Horney discusses other contradictions within society that contribute to neurosis, and they can help us to understand our contemporary problems:

> The first contradiction to be mentioned is that between competition and success on the one hand, and brotherly love and humility on the other. On the one hand everything is done to spur us toward success, which means that we must be not only assertive but aggressive, able to push others out of the way. On the other hand we are deeply imbued with Christian ideals which declare that it is selfish to want anything for ourselves, that we should be humble, turn the other cheek, be yielding. For this contradiction there are only two solutions within the normal range: to take one of these strivings seriously and discard the other; or to take both seriously with the result that the individual is seriously inhibited in both directions. (Horney, 1937: 287)

Horney's first contradiction is between two "cultural values," or personal values that have come to be widely shared because of society's impact on the individual. One of them, which has been labeled within sociology as "achievement and success" (Williams, 1970: 452–500), illustrates what might be called a work-related cultural value. Illustrations of other work-related cultural values throughout contemporary societies are "activity and work," "economic progress," "material comfort," and "efficiency and practicality." By contrast, "love" is part of the more general cultural value of "close relationships," which in turn illustrates the still more general category, or people-oriented cultural values. Other people-oriented cultural values are equality, freedom, democracy, and individual worth. Our one-sided development of physical science has yielded an overriding emphasis on materialism, by contrast with people-oriented cultural values, yielding the contradiction that Horney sees as creating neurosis in the individual.

Horney also presents her approach to confronting the problem posed by the first contradiction—that between "competition and success on the one hand, and brotherly love and humility on the other": "For this contradiction there are only two solutions within the normal range: to take one of these strivings seriously and discard the other; or to take both seriously with the result that the individual is seriously inhibited in both directions" (Horney, 1937).

Her conclusion is similar to Gautama's solution for the problem of dukkha, as discussed in chapter 2: abandon one of these cultural values, just as Gautama suggested that we must remove our "thirst" or aspirations, or else become the victim of neurosis.

Horney centers on the problem of neurosis, a term that was popular in her time but has gone out of fashion even among psychiatrists. A sociological concept centering on the emotions of the individual, one that is linked to other social science concepts and is popular as well, is "alienation." It was Karl Marx in his early writings who is best known for advancing this concept:

> We have now considered the act of alienation of practical human activity, labour, from two aspects: (1) the relationship of the worker to the product of labour as an alien object which dominates him ... (2) the relationship of labour to the act of production within labour. This is the relationship of the worker to his own activity as something alien and not belonging to him.... This is self-alienation as against the above-mentioned alienation of the thing.... Since alienated labour: (1) alienates nature from man; and (2) alienates man from himself, from his own active function, his life activity; so it alienates him from (3) the species.... For labour, life activity, productive life, now appear to man only as means for the satisfaction of a need, the need to maintain his physical existence ... free, conscious activity is the species-character of human beings. (4) A direct consequence of the alienation of man from the product of his labour, from his life activity and from his species-life, is that man is alienated from other men. (Marx, 1844/1964: 125–127, 129)

Marx's breadth is most refreshing when we compare him to contemporary specialists and subspecialists throughout the social sciences. His four aspects of alienation focus on physical structure, personality structure, biological structure, and social structure. Following Marx, we might see alienation as persisting feelings of isolation from self, others, one's own biological structure, and the physical universe. More concretely, the worker is dehumanized by repetitive activities (biological structure), he has no control over those activities (personality structure), he is divorced from the product of his work (physical structure), and he is also separated from his fellow workers (social structure). For Marx, alienation is closely linked to the patterns of social stratification or persisting hierarchy between owners and workers in the workplace, all tied to a capitalistic economic system. Marx also had a long-term sense of history. He saw all of history as involving class conflicts, such as that between owners and workers. And he believed that the situation of the worker was continuing to worsen:

> The more the worker produces the less he has to consume; the more value he creates the more worthless he becomes; the more refined his product the more crude and misshapen the worker; the more civilized the product the more barbarous the worker; the more powerful the work the more feeble the worker; the more the work manifests intelligence the more the worker declines in intelligence and becomes a slave of nature. (Marx, 1844/1964: 122)

Marx's analysis, in relation to Figure I-1, suggests that the bottom curve of the fulfillment of aspirations does not move up gradually but rather moves down: there is less and less fulfillment of aspirations as history proceeds. As for the top curve of aspirations, Marx had nothing to say about it. This neglect of cultural values was one limitation of his work, which focused on relationships like patterns of social stratification.

Georg Simmel, one of the founders of sociology who wrote half a century later, gives us further insight into the emotional problems of the individual as industrialization proceeded. His approach focused on culture throughout society as well as on the situation of all individuals, not just workers:

> The deepest problems of modern life flow from the attempt of the individual to maintain the independence and individuality of his existence against the sovereign powers of society, against the weight of the historical heritage and the external culture and technique of life.... the essentially intellectualistic character of the mental life of the metropolis becomes intelligible as over against that of the small town which rests more on feelings and emotional relationships.... the metropolitan type ... creates a protective organ for itself against the profound disruption with which the fluctuations and discontinuities of the external milieu threaten it. Instead of reacting emotionally, the metropolitan type reacts primarily

> in a rational manner, thus creating a mental predominance through the intensification of consciousness. (Simmel, 1903/1971: 324–325)

In common with Marx, Simmel was oriented to the importance of history and, as a result, to Timeland. Also in common with Marx, Simmel was oriented to a very broad understanding of human behavior, pointing him toward Spaceland as well. But by contrast with Marx, Simmel's interest in culture along with the full range of individuals in society—and not just a focus on workers in the workplace—was broader than Marx's interest. Simmel was more scientific in avoiding beliefs in the "inevitability" of any given future event. Further, Simmel's analysis of the individual in all of his or her uniqueness was most profound. Not many social scientists have discussed individual emotions—granting that Freud, neo-Freudians, and modern psychotherapists discovered their centrality in understanding the problems of the individual—yet they were most important for Simmel. Just as Freud and Horney analyzed emotional repression, so did Simmel see this happening in the metropolis. For Simmel, this affected everyone in the city and not just some individuals who happened to have unique experiences. By so doing, he joined Horney with her focus on contradictions between cultural values as well as between the encouragement of our aspirations and our ability to fulfill them.

We can take a very long and general view of Simmel's analysis here, following a Timeland and Spaceland orientation. Is it in fact true that our industrial and technological developments over the past five centuries have yielded contemporary societies where we have all learned—more or less—to bury our deepest emotions? Simmel argued that this was indeed true for the inhabitants of large cities in his time. He saw people dealing with others just as they dealt with numbers—in an intellectual by contrast with an emotional fashion: "These relationships stand in distinct contrast with the nature of the smaller circle in which the ... knowledge of individual characteristics produces ... an emotional tone in conduct, a sphere which is beyond the mere objective weighting of tasks performed and payments made" (Simmel, 1903/1971: 327). From this perspective, we can understand more fully just why we all—and not just the Springdalers or those individuals living in large cities—have great difficulty in becoming aware of the aspirations-fulfillment gap in our own lives, or why we prisoners in Plato's cave are content to see shadows on the wall rather than learn to look at our own personal problems. And we can also understand more clearly why Pogo's statement, "We have met the enemy and he is us," is so appropriate as an implicit indictment of the outward-orientation of modern society.

Are we moderns emotional cripples who have great difficulty not only understanding our own feelings but also reaching out to others? If so, here is yet another huge gap between our aspirations and the reality. Elie Wiesel, the American writer, Nobel Prize winner, and Auschwitz survivor has claimed the following:

> The opposite of love is not hate, it's indifference.
> The opposite of art is not ugliness, it's indifference.
> The opposite of faith is not heresy, it's indifference.
> The opposite of life is not death, it's indifference. (quoted in Knowles, 2004)

For Wiesel, indifference is a fundamental failure of the human race to fulfill the nature of its humanity. It was indifference to the fate of six million Jews that helped to make the Holocaust possible. It is indifference to our own fate that helps us today to bury our most pressing problems, using the Springdalers's techniques of particularization and falsification of memory that keep us in Flatland rather than in Spaceland and Timeland. Greek civilization called for a balance of "head," "heart," and "hand"; yet it appears that contemporary society has taken us away from that balance. Max Weber, another founder of the discipline of sociology along with Marx and Simmel, commented on the negative impact of modern achievements in the closing paragraphs of *The Protestant Ethic and the Spirit of Capitalism*: For of the last stage of this cultural development, it might well be truly said: "Specialists without spirit, sensualists without heart; this nullity imagines that it has attained a level of civilization never before achieved" (1905/1958: 182).

Some social scientists had argued in the 1960s and 1970s that modern procedures of automation would change this situation and reduce alienation:

> There have been many expressions of hope in the last two or three decades that the coming of automation would reduce the amount of alienation in the modern workplace by replacing labor of the most mechanical and mindless kind with activities that require skill, judgment, and a sense of craft. Workers in charge of automated equipment, so the argument goes, relying on quick intelligence and sure perceptions rather than on brute strength are freed from the unrelenting rhythms of a machine, can wander around the larger workplace and develop some feeling for what the whole enterprise is about, and, in general, escape from the pinched and narrow niches into which a complicated division of labor would otherwise have confined them.
>
> Unfortunately, this hope has not borne fruit. Automation processes apparently do not require the hoped-for "skill, judgment, and a sense of craft." They work to shift control from the worker to the worker's superiors, who make the decisions that guide the work process, thus strengthening patterns of social stratification. Also, automation appears to replace one kind of boredom with another kind. Instead of endlessly repeating a certain kind of muscular activity, the worker in an automated factory or continuous-process plant is confronted by the experience of doing nothing at all. Further, the worker is moved far away from the actual materials that the machines are processing, by contrast with the craftsworker who feels the grain of the wood or the weave of the fabric.

> In addition, automated processes create night shifts for a large proportion of the labor force, taking them away from their relationships with others during the day, when they must sleep. (Erikson, 1986: 4–5)

If we focus on one emotion that appears to have special importance in contemporary society—shame or embarrassment—then we can gain further insight into problems such as the increasing aspirations-fulfillment gap along with our commitment to a stratified worldview. Thomas J. Scheff, whose work has been central to the development of the Web and Part/Whole Approach to the scientific method (1990, 1994, 1997), has viewed shame in historical perspective:

> The Old Testament contains many, many references to pride and shame but very few to guilt. The New Testament reverses the balance.... It is possible that the role of shame in social control has not decreased but has gone underground instead.... For example, we say, "It was an awkward moment for me.... It contains two movements that disguise emotion: denial of inner feeling and projection of it onto the outer world. I was not embarrassed; it was the moment that was awkward. (quoted in Phillips 2001: 151)

Scheff, together with his colleague and wife, Suzanne Retzinger (1991), have come to see hidden versus acknowledged shame as fundamental to a range of modern problems. And they see hidden shame as increasing along with the industrialization and modernization of the past centuries. Yet why has this happened? If we refer to an increasing aspirations-fulfillment gap as we continue to modernize, then we can begin to understand the increasing repression of shame. If we had a direction for closing that gap, such as that offered by a broad approach to the scientific method, then there would be no need to repress our shame. Without such a direction, however, we have little choice. Instead of acknowledging our failures to ourselves and others, we bury our shame much like the Springdalers. And if that gap is increasing, then we will be constrained to increase our efforts to repress our shame at being unable to narrow that gap.

It appears, then, that not only are the problems of the individual increasing as modernization is proceeding, but, in addition, we are escaping more and more from efforts to confront those problems. Following Karen Horney's analysis, we are becoming ever more neurotic as advertising pushes us toward wanting more and more that we are unable to buy. If we turn to Marx, we find an argument for the worsening of the worker's situation or the decreasing ability to fulfill his or her wants. Simmel argues that we are losing our ability to express our emotions and reach out to others in modern society. Weber claims in a similar vein that we are becoming "Specialists without spirit, sensualists without heart." And Scheff and Retzinger give us evidence that we are learning to bury key emotions like shame instead of learning to confront our problems. Apparently we are retreating from the ideals and optimism of the Renaissance and the Enlightenment, which emphasized the human being's possibilities. If we continue to travel on this path toward pessimism and helplessness, there is little question as to what will happen in the future.

Erich Fromm, a neo-Freudian who was interested in culture and human possibilities no less than Karen Horney, wrote in *The Sane Society* (Fromm, 1955) that "the danger of the past was that men became slaves. The danger of the future is that men may become robots." For if, indeed, we continue to bury our emotions as our problems increase without any visible alternative to confront them, our behavior will become more and more robotlike. Should we look forward to this as our future? Or is there an alternative that we can adopt?

Expressive Orientation

> We shall not flag or fail. We shall go on to the end. We shall fight in France, we shall fight on the seas and oceans, we shall fight with growing confidence and growing strength in the air, we shall defend our island, whatever the cost may be. We shall fight on the beaches, we shall fight on the landing grounds, we shall fight in the fields and in the streets, we shall fight in the hills; we shall never surrender.
>
> —*Winston Churchill, speech in the House of Commons, June 4, 1940*

If ever there were a statement embodying the complete emotional commitment of an individual to an extremely broad, vital, and difficult task, it was Churchill's speech on June 4, 1940, a date—although in a different year—that happens to be my birthday. Churchill was faced with an extremely strong, committed, and ruthless enemy; and since the United States was not yet ready to fully commit itself, he was going it alone. This was a statement contradicting two lines in Yeats's "The Second Coming" that were not quoted in chapter 1:

> The best lack all conviction, while the worst
> Are full of passionate intensity.

I interpret "the worst" who are full of passionate intensity as fanatics like Hitler, who were committed to a very narrow agenda, versus a broad and humanistic approach. "The best," with humanistic breadth, are aware of complexity and contradictions and, thus, generally remain unable to act forcefully. Churchill is most unusual in his combination of breadth and forcefulness. Churchill's deep commitment to a problem is much the same as the commitment required by the first step of a broad scientific method. Such emotional commitment makes all the difference between a sustained and effective scientific effort and one that achieves little and has negligible impact.

It is important to build on the ideas in chapter 2 as we proceed with chapter 3. In particular, let us return to the poetry of Walt Whitman. Given the widespread pessimism in our own age, we desperately require optimism, and Whitman's poetry begins to fulfill this need. The following quote is from his "Song of Myself," just as was the earlier quote:

> This is the meal equally set, this the meat for natural hunger,
> It is for the wicked just the same as the righteous, I make appointments with all,
> I will not have a single person slighted or left away....
>
> Divine am I inside and out, and I make holy whatever I touch or am touch'd from,
> The scent of these arm-pits aroma finer than prayer,
> This head more than churches, bibles, and all the creeds....
>
> I think I could turn and live with animals, they are so placid and self- contain'd,
> I stand and look at them long and long.
> They do not sweat and whine about their condition,
> They do not lie awake in the dark and weep for their sins,
> They do not make me sick discussing their duty to God,
> Not one is dissatisfied, not one is demented with the mania of owning things,
> Not one kneels to another, nor to his kind that lived thousands of years ago,
> Not one is respectable or unhappy over the whole earth.
> (1892/2004: 38, 44, 50)

Whitman deeply believed in the spirit of the evolutionary worldview with its focus on the individual's unbelievable possibilities. He sets a meal for the wicked as well as the righteous, for the wicked are human no less than the righteous, and they still have the possibility of developing their human potential. His fundamental belief in the ultimate worth of all human beings shines through. For Whitman, the human being is nothing less than a divine creature, and we should learn to worship ourselves "more than churches, bibles, and all the creeds." And we have much to learn from the animals, for "they are so placid and self-contain'd," and "Not one is dissatisfied, not one is demented with the mania of owning things, Not one kneels to another." The unhappiness of human beings, by contrast with the animals, is illustrated by our aspirations-fulfillment gap. Let us recall here the grocer of Balgat from the introduction who said, "I have not the possibility in myself to get the things that I want. They only bother me." Following the idea of a stratified worldview, much unhappiness derives from our patterns of social stratification. For those patterns get in the way of our using a broad approach to the scientific method in our everyday lives. By kneeling to others, by contrast, we lose our freedom to focus on our own personal experiences as a basis for confronting and solving our problems. Here again Whitman reveals his distaste for a stratified worldview and his interest in an evolutionary one.

If in the first section of this chapter, I opened a door to the importance of Freud and Marx for understanding how we can use the scientific method to change our worldview; then in this section I wish to open the door to the philosophy of existentialism and, in particular, to the work of two philosophers: Friedrich Nietzsche and Jean-Paul Sartre. It is no accident that it is to philosophy that we turn here, for it is philosophers who have attempted to reach as deeply as possible into our fundamental assumptions, our metaphysical stance, or our worldview as a basis for understanding ourselves and the world. For too long social scientists have proceeded to give no more than lip service to the importance of the history of philosophy. After all, many would claim, was it not philosophers during Europe's Dark Ages that kept would-be scientists from testing ideas against experience, versus bowing down to authority? Yet it was also philosophers like Nietzsche and Sartre who gave us much of what we know about the importance of not bowing down to authority through the examples of their own accomplishments.

Nietzsche, whose importance is widely misunderstood because some of his work was subsequently used by Hitler to help justify his own efforts at world conquest, has given us—among many other ideas—the idea of the death of God:

> Have you not heard of that madman who lit a lantern in the bright morning hours, ran to the market place, and cried incessantly: "I seek God! I seek God!" ... The madman jumped into their midst and pierced them with his eyes. "Whither is God?" he cried; "I will tell you. We have killed him—you and I. All of us are his murderers. But how did we do this? How could we drink up the sea? Who gave us the sponge to wipe away the entire horizon? What were we doing when we unchained this earth from its sun? Whither is it moving now? Whither are we moving? Away from all suns? Are we not plunging continually? Backward, sideward, forward, in all directions? Is there still any up or down? ... Here the madman fell silent and looked again at his listeners; and they, too, were silent and stared at him in astonishment. At last he threw his lantern on the ground, and it broke into pieces and went out. "I have come too early," he said then; "my time is not yet. This tremendous event is still on its way, still wandering; it has not yet reached the ears of men. Lightning and thunder require time; the light of the stars requires time; deeds, though done, still require time to be seen and heard. (1887/1974: 181–182, #125)

Nietzsche was a philosopher who wrote in a literary vein, emphasizing the importance of figures of speech like "the death of God." My own interpretation is that he was referring to the loosening of the hold that social stratification has on all of us, as illustrated by our stratified worldview or metaphysical stance. Of course, many have interpreted it to focus on its meaning for religions. Yet from my own perspective that meaning is primarily its orientation to the persisting hierarchies within religions. To discuss "the death of God," then, is nothing less than discussing a change in our worldview or metaphysical stance. Yet this is so fundamental a change in society and the individual that Nietzsche must use figures of speech to communicate its importance. For it is figures that communicate directly with senses such as sight and sound, as illustrated by this passage: "How could we drink up the sea? Who gave us the sponge to wipe away the entire horizon? What were we doing when we unchained this earth from its sun?" Nietzsche is, then, a most unusual philosopher, someone who combined "head" and "heart" in his efforts to reach his audience.

Yet let us note that the madman—and he was certainly mad in the sense that he was violating absolutely fundamental traditions—noted that his listeners remained silent and stared at him in astonishment. As a result, he concluded, "I have come too early . . . my time is not yet." We have begun to question the social stratification and stratified worldview that dominates our lives, given the Renaissance along with the American and French revolutions with their emphasis on equality and the worth of the individual. Yet this is no more than a beginning. We have as yet a great deal to do to achieve the death of the God of social stratification and a stratified worldview. For one thing, we require a very broad scientific method that we all can use in our everyday lives. For another thing, we require an alternative worldview or metaphysical stance that promises to help us confront our escalating problems. And for still another thing, we must learn to actually use that broader scientific method and that alternative worldview in our own lives, taking one step at a time to actually achieve "the death of God."

As for how to develop such a scientific method, Nietzsche is able to give us some advice:

> *Taking seriously*—In the great majority, the intellect is a clumsy, gloomy, creaking machine that is difficult to start. They call it "taking the matter *seriously*" when they want to work with this machine and think well. How burdensome they must find good thinking! The lovely human beast always seems to lose its good spirits when it thinks well; it becomes "serious." And "where laughter and gaiety are found, thinking does not amount to anything": that is the prejudice of this serious beast against all "gay science."—Well then, let us prove that this is a prejudice. (1887/1974: 257, #327)

Of course, "gay" for Nietzsche writing in his time had the connotation of "joy" or "gaiety" versus seriousness. For Nietzsche, the scientific method should by no means be "a clumsy, gloomy, creaking machine that is difficult to start." Rather, "good thinking" accompanies such gaiety, and not gloom. Nietzsche is, thus, giving us a formula for a scientific method that all of us can use most easily as we proceed with our lives, a formula that we will be glad to use because it adds joy to our behavior. Given the high degree of specialization and hierarchy we have developed in the modern world, we have learned to take the scientific method away from ordinary folk and place it in the hands of a scientific elite, an elite who themselves have not learned to use it in their everyday lives. Yet Nietzsche is pointing toward a kind of scientific method that all of us can use from one moment to the next, one that adds joy rather than gloom to our lives.

Nietzsche has more to tell us about what we have done to move toward such a scientific method, implying what we must do in the future:

> *On the doctrine of poisons*—So many things have to come together for scientific thinking to originate; and all these necessary strengths had to be invented, practiced, and cultivated separately. As long as they were still separate, however, they frequently had an altogether different effect than they do now that they are integrated into scientific thinking and hold each other in check. Their effect was that of poisons; for example that of the impulse to doubt, to negate, to wait, to collect, to dissolve. Many hecatombs of human beings were sacrificed before these impulses learned to comprehend their coexistence and so feel that they were all functions of one organizing force within one human being. And even now the time seems remote when artistic energies and the practical wisdom of life will join with scientific thinking to form a higher organic system in relation to which scholars, physicians, artists, and legislators—as we know them at present—would have to look like paltry relics of ancient times. (1887/1974: 173, #114)

For Nietzsche, the scientific method is indeed a complicated beast. Its development has required many separate developments that had to be integrated if that method were to emerge in modern times. While each of those developments remained separated from the others, however, they acted as "poisons," as illustrated by "the impulse to doubt, to negate." That impulse yields a sense of problem, but as we have seen—especially following our pendulum metaphor—that a sense of problem must be combined with a swing of the scientific pendulum in the direction of making progress in solving the problem. Otherwise that sense of problem ends in pessimism rather than the optimism that we have learned is so characteristic of the scientific method. Nietzsche then proceeds to illustrate the swing of the scientific pendulum at the end of his passage, when he refers to the failures of scientific thinking to have come far enough to integrate the experiences of "scholars, physicians, artists, and legislators." In other words, we still require a much broader approach to the scientific method, and this is the case in our own time even more than in Nietzsche's time, given the problems that we are facing at this time in history.

Nietzsche was an early "existentialist," and we turn to Jean-Paul Sartre—as explained by the American philosopher, Abraham Kaplan—to learn more about the nature of existentialism.

In *The Brothers Karamazov* Dostoevsky (another forerunner of existentialism) has two characters engage in one of his characteristically interminable yet endlessly fascinating discussions about the existence of God; and one of them—I think Ivan—concludes: "But you see, if there were no God, everything would be possible!" The existentialist is a philosopher who approaches the basic problem of human existence in just these terms, that indeed, everything *is* possible. The problem then becomes to decide which among the infinite possibilities is to be actualized by human choice. On what basis can such a choice be made?

> Only man exists in the sense that what he is is not limited, specified, determined beforehand—in a word, defined prior to his existence.... The first principle of existentialism then, as Sartre formulates it, is this: "Man is nothing else but what he makes of himself." Everything else in the world is only what has been made of it by the rest of being.... Now to conceive of man as the existent which determines its own essence is to recognize that the most fundamental attribute of the human being is his capacity for choice. (Kaplan, 1961: 101, 103-104)

For Sartre, the human being "exists," by contrast with everything else in the universe that is an "essence." In other words, anything and everything is possible for the human being, since he or she cannot be labeled once and for all as this "essence" or that "essence." We humans have the potential to choose from one moment to the next, and we can change whatever it is that we were or were thought to be. Thus, man makes himself or exists, whereas all other things are essences determined by forces outside their control. For man, all things are possible. We can see in Nietzsche's ideas the tremendous optimism that this belief can yield, granting the despair felt by most existentialists about the human situation. If everything is possible, then we need not despair about the present situation of the human race. We can learn to change our situation. For example, we can learn to develop and use in everyday life an extremely broad and powerful scientific method. We can also learn to change our stratified worldview. And we can learn to apply a new worldview in confronting and making progress on our fundamental problems. Further, we can continue to reinvent ourselves so long as we live. We are the only organisms who have this capacity, this freedom.

However, if the actions we take are merely to conform to the wishes of others, then we are not making "genuine" or "authentic" choices. We are avoiding our fundamental responsibility to ourselves and to society. In a deep sense, we are failing to act as human beings with our own unique personalities. We are escaping from the freedom that we have. Sartre does not mean that we must make only those choices that are different from the choices that everyone else makes. Rather, our choices should stem from whoever we are as individuals. Shakespeare had something to say about this in *Hamlet* when Polonius gives advice to his son, Laertes, before Laertes' journey:

> This above all: to thine own self be true,
> And it must follow, as the night the day,
> Thou canst not then be false to any man.
>
> —*Shakespeare,* Hamlet, *1601, Act 1, Scene 3, Line 78*

What existentialism achieves is a clear focus on the possibilities for personal evolution of every single individual. Its focus is on the upper curve of Figure I-1: our aspirations should know no limit. Yet we have found that unless aspirations are accompanied by the fulfillment of those aspirations, the result is an increasing aspirations-fulfillment gap. And that gap in turn has severely negative repercussions. For one thing, it encourages the individual—like the Springdaler—to bury any awareness of problems, as further illustrated by Simmel's analysis of the metropolis and Scheff's analysis of shame in contemporary society. This in turn gets in the way of confronting problems like those uncovered by Marx and Horney, such as the problem of alienation along with contradictions between cultural values. Nevertheless, in these contemporary times of increasing social problems when pessimism about human possibilities is so widespread, we desperately need the optimism about human possibilities that existentialist thinking offers.

Existentialists themselves generally despaired about the human situation, but that despair was based on their failing to see an effective and concrete way for the individual to learn how in fact to make authentic or genuine choices. That has to do with the lower curve of Figure I-1. If the individual can learn to fulfill increasing aspirations one step at a time by using a very broad scientific method in everyday life, then his or her aspirations can continue to climb without any limit. This can occur, I believe, if the individual also learns to change from a stratified to an evolutionary worldview so as to encourage that broader scientific method.

It is useful to update existentialist ideas by referring to a figure who is neither identified with existentialism nor identifies himself with that movement, yet who carries forward key existentialist ideas: Phillip C. McGraw or "Dr. Phil" as he is widely known. This passage comes from his *Self Matters*:

> Let me tell you what I would wish for you to be thinking and saying now, during, and after you read this book: Hey, wait a minute here. Screw the expectancies; screw living for everyone else.... I want to make *me* happy by being true to myself doing what I care about. If I love music, I want to have music in my life. If I want a career, then I want to find a way to have

> it. If I'm tired of being fat, I want to prioritize that change into my lifestyle. If I'm not being treated with dignity and respect, that's not okay, not now, not ever. I would rather be alone than sick with someone else.... I'm tired of being scared all the time. Scared about kids, money, job, boss, parents, and acceptance ... I want my kids to know and have all of me instead of some half-assed counterfeit.... I want to live with peace, fulfillment, joy, and excitement.... I want to feel calm and peaceful. I want to feel satisfied. I want to be able to say, "I feel good." (McGraw, 2001: 18-19)

McGraw illustrates in this passage a central idea from existentialism: the human being's possibilities for creating self and world. Man makes himself, and we humans have the capacity to change our situation if we choose to do so. Everything is indeed possible, and we need not settle for a life that we do not want to continue to lead. What McGraw achieves in this quote is to communicate a number of gaps between what people want and what they are in fact getting. That is the opposite of the Springdalers's pattern of burying those gaps with techniques of particularization and the falsification of memory. By contrast, McGraw taps into the reader's awareness of fundamental problems in his or her life, such as problems with one's weight, with one's career, with being treated disrespectfully, with fear, with lack of joy, with lack of excitement, and with a sense of dissatisfaction. He uses the same approach to generate the reader's motivation that I use in this book in the first half of each chapter. McGraw reveals gaps between what we want and what we get, such as our deep sense of alienation—following Marx, Horney, and Simmel—versus our desire to feel good about ourselves, our relationships with others, and the world in general. In this way, McGraw is following a broad approach to the scientific method, where one starts with an awareness of and commitment to a problem.

In his book McGraw also illustrates other aspects of a broad scientific method. For one thing, he is reflexive to a degree, giving personal examples of problems that he at first avoided in his life and then proceeded to confront and solve. He also gives many concrete illustrations of his experiences as a psychotherapist. Thus, he is able to move both up and far down language's ladder of abstraction. He moves up to basic problems that people generally have, such as lack of "peace, fulfillment, joy, and excitement." Then he moves far down to examples of individuals he has treated who have made progress in solving such problems. Further, his procedures emphasize the importance for the reader and the client to become reflexive: to go back over the details of one's own life and think of situations that illustrate what one no longer wants as well as what one would like to have. These are procedures for taking seriously what Socrates told us in the fourth century B.C.: "The unexamined life is not worth living," and his advice to "know thyself." He provides specific directions for solutions, by contrast with Sartre.

Yet McGraw, like Sartre, fails to build on what social scientists have discovered about the problems that the individual encounters when attempting to act in an "authentic" or "genuine" way to solve personal and world problems. Sartre and McGraw emphasize the top curve of Figure I-1 with precious little to help the individual raise that bottom curve so that it approaches the top one. His specific advice to analyze one's own past is helpful, but it lacks insight into the powerful yet largely invisible forces operating on the individual, forces that social science concepts can detect. Although a number of individuals are able to find ways to raise that bottom curve on their own or with the aid of McGraw's advice, for many others McGraw's assumption—like Sartre's—that the individual can be successful without knowledge of what the social sciences have discovered serves only to increase their aspirations-fulfillment gap. It is easy to understand McGraw's failure to build on social science knowledge, however, once we take into account the failure of social scientists themselves to integrate available knowledge about human behavior and problems. Nevertheless, McGraw and Sartre do succeed in communicating a highly optimistic view of human possibilities, something sorely needed in these pessimistic times.

As we proceed from chapter 3 to chapter 4, we move from an emphasis on the heart to a focus on the "head." In particular, considerable attention is devoted to the nature of our worldview or metaphysical stance, an idea that is extremely abstract yet extremely important. In my view, it is our stratified worldview that is holding back our ability to develop a broad scientific method and use that method in our everyday lives. For that worldview is narrow and static, by contrast with our ideals for the scientific method. And our worldview is sufficiently powerful to trump our epistemology, or our understanding of how to learn or achieve understanding of any given phenomenon. Nevertheless, our use of a broad scientific method can indeed yield—little by little—changes in our worldview, despite the latter's enormous power. We have gained at least the beginnings of an alternative worldview, and this makes it possible to confront a worldview that is preventing us from addressing the threatening problems of our era. Following Nietzsche, the God of our stratified worldview is dead, but "this tremendous event is still on its way, still wandering; it has not yet reached the ears of men. Lightning and thunder require time; the light of the stars requires time; deeds, though done, still require time to be seen and heard" (1887/1974). I would add that the death of the God of stratification requires not only time but also our committed efforts, paralleling the commitment of Winston Churchill.

CHAPTER 4

"Head"

Stratified or Bureaucratic Versus Evolutionary Worldview

ORDINARY OR EVERYDAY LANGUAGE IS THE MOST IMPORTANT PHENOMENON distinguishing us humans from all other forms of life. It enables us to solve problems in any given situation by reaching out to other situations in time and space. As a result, it enables all of us to use a method for solving problems that is similar to the scientific method by pulling together information relevant to a given problem and helping us decide how to solve it. George A. Kelly, a clinical and academic psychologist, carried this idea forward in the middle of the twentieth century:

> To a large degree—though not entirely—the blueprint of human progress has been given the label of "science." Let us then, instead of occupying ourselves with *man-the-biological organism* or *man-the-lucky-guy,* have a look at *man-the-scientist*. . . . every man is, in his own particular way, a scientist. . . . Might not the individual man . . . assume more of the stature of a scientist, ever seeking to predict and control the course of events with which he is involved? Would he not have his theories, test his hypotheses, and weight his experimental evidence? (1963: 4)

Following Kelly, it appears useful for all of us to see ourselves to a degree as scientists, for in general we do go about solving problems in a similar way. Let us also recall from chapter 1, however, Gouldner's reference to the limitations of "everyday language and understanding," and his idea that they can be transcended by the "extraordinary languages" of the social sciences. For that extraordinary language to help us move beyond "the deficient understandings of ordinary language," however, it must enable us to integrate a great deal of information. This was illustrated in the introduction by the concept of "social stratification," which is sufficiently general so as to include persisting hierarchies invoked by such concepts as "sexism," "nationalism," "racism," "ethnocentrism," "religious prejudice," and "classism." By using the concept of social stratification together with the other general concepts in the chapter titles of this book, we begin to learn the extraordinary language of the social sciences. And by so doing we begin to learn how to use a very broad approach to the scientific method in our everyday lives. This approach by no means substitutes for the everyday language we are already using habitually, a language that helps us to solve our everyday problems. Rather, by opening us up to a much wider range of information it enables us to develop much further our problem-solving abilities. Yet for this to occur we must learn to make that extraordinary language a part of our everyday thought and speech no less than the part played by our ordinary language. Still further, to solidify the role of that extraordinary language within our problem-solving efforts, it would also be useful—following Kelly—for us to come to see ourselves as scientists.

Figure 4-1, "Individual Structures," portrays relationships among most of the key concepts within this extraordinary language, which we have already examined, together with several additional concepts put forward in this chapter and chapter 5. I will limit any presentation of specific definitions in the interest of brevity, although definitions are available in the glossary following chapter 11. By contrast with our ordinary language, these concepts of extraordinary language relate to one another in a highly systematic way. Also by contrast with ordinary language, these concepts are very general and, as a result, are broad enough to serve as bridges linking ordinary-language concepts, just as "social stratification" links "sexism," "nationalism," "racism," "ethnocentrism," "religious prejudice," and "classism."

Before examining the extraordinary concepts in Figure 4-1, let us not forget the importance of our learning to link them with our ordinary concepts and using them in everyday life. For example, "beliefs" are similar to ordinary concepts like "ideas," "opinions," "views," "conclusions," and "convictions." Each of these everyday concepts has a different shade of meaning, and each one has proved to be of some value in yielding understanding of an individual's beliefs. In addition, however, it is valuable to use a concept such as "beliefs" that can serve as an umbrella to include all of these concepts, just as it is valuable to use the concept of "flower" to include specific concepts like "rose," "daffodil," and "daisy."

In this way, these extraordinary concepts can help us to move far out into the dimensions of space and time or, following the figures of speech presented in chapter 1, into Spaceland and Timeland. As for space, the general concept "beliefs" encompasses all of the ideas that an individual has learned about self and world and, thus, invokes an entire universe of phenomena. As for time, "beliefs" are structures in the sense that social scientists view them as persisting over time, by contrast with "opinions," which change more easily as new information is gained. This breadth in space and time is also the basis for definitions of all of the other concepts in Figure 4-1.

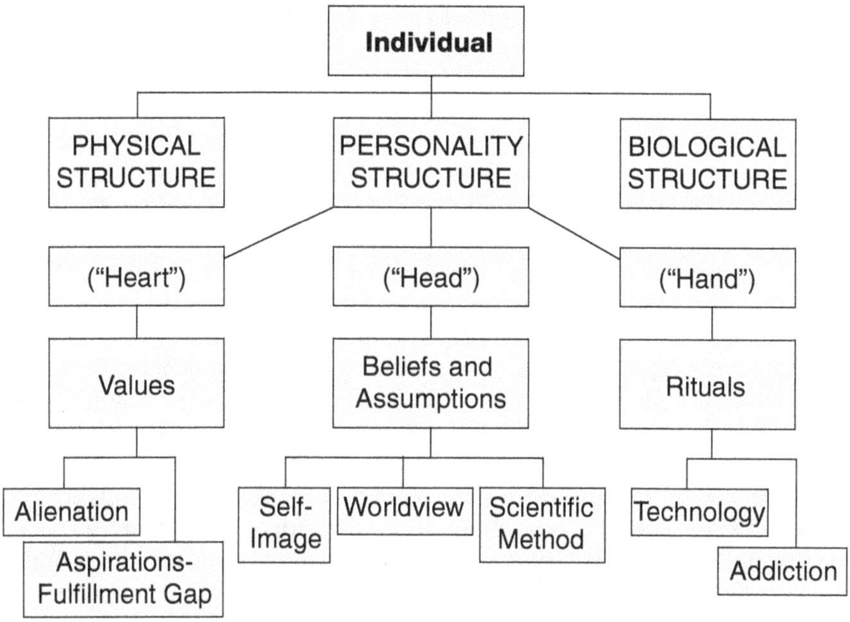

Figure 4-1. Individual Structures

Many ordinary concepts also reach out very far in space and time, such as "universe" and "millennium." Yet my concern is not just with Spaceland and Timeland characteristics, but also with the ability to link concepts to one another in a highly systematic way so as to provide a framework for integrating our full range of ordinary concepts. For that purpose, "physical structure" is far more appropriate, given its linkage with all physical phenomena as well as its persistence within the time dimension.

Before examining further the specific concepts in Figure 4-1, it is important to note that there are a great many other concepts that social scientists have made good use of in their efforts to explain the nature of the individual. My purpose here is by no means to exclude those concepts, just as my purpose is not to exclude the ordinary concepts that we use in everyday life. Rather, it is to provide general concepts that function like umbrellas that can include a great many other specific concepts. Further, there is no hard-and-fast agreement among social scientists that the concepts depicted in Figure 4-1 are in fact the best ones to convey the general nature of the individual, based on social science research. I selected them because I believe that they are the best ones for integrating my own understanding of the individual, especially given the ease of linking them in a systematic way. Of course, my own understanding will change over time as I learn more about the nature of the individual. This situation contrasts with that in the physical sciences, where the simpler phenomena involved have yielded greater agreement about the key concepts required to understand physical structures.

As for the concepts depicted in Figure 4-1, it was in chapters 1 and 2 that we centered on "physical structures" and "biological structures," concluding that our consideration of these structures is essential for achieving an understanding of the "individual." For example, just now I have been contrasting language that takes us far out into the physical dimensions of space and time with language that fails to do so. In chapter 2 we noted the importance within human behavior of perception, as illustrated by the power of advertising on television. "Personality structure" is the umbrella concept I am using for the many additional structures within the individual that we examine in chapters 3, 4, and 5. The figures of speech of "heart," "head," and "hand" are useful in part because they invoke our sense of vision and, thus, are linked to our biological patterns of perception. They are also useful because they alert us to the breadth of the human personality with its inclusion of emotions or feelings along with ideas or opinions as well as actions and interactions. We shall become aware of the further usefulness of these figures when we look to chapters 6–8 on social structures and chapters 9-11 on the situation, as suggested in those chapter titles.

We have already begun to examine the concept of "beliefs." Let us note that it is paired with "assumptions." The difference between the two has to do with an individual's degree of awareness or consciousness and assumptions being closer to the pole of consciousness and awareness being closer to the pole of a lack of consciousness or awareness. This distinction is most important, given what has been learned about the power of unconscious ideas—or assumptions—in shaping our behavior, as illustrated by the concepts "self-image," "scientific method," and "worldview." We do indeed have some awareness of each of these phenomena, although we act on many assumptions bearing on all three concepts that we remain unaware of. The importance of the concept of "worldview" has already been indicated in the introduction as well as in each of the foregoing chapters. For it is our learning how to change from a stratified to an evolutionary worldview—with the aid of the scientific method in our everyday lives—that is fundamental to narrowing the aspirations-fulfillment gap and solving personal and world problems. The idea of self-image is tied closely to a reflexive orientation, which is basic to the Web and Part/Whole Approach to the scientific method. Also, social science research indicates the centrality of one's view of oneself for understanding one's behavior.

Moving from "head" to "heart," our "values" include our most general and persisting goals (such as achieving happiness and living a meaningful life) along with the more specific yet persisting goals of a given individual (such as traveling around the globe). "Alienation"—as we have seen in chapter 3—is tied to one's lack of fulfillment of important values that are widely shared, such as achieving close relationships with others or doing work that is meaningful. "Aspirations-Fulfillment Gap" refers to an individual's values-fulfillment or aspirations-fulfillment, also discussed in the introduction and the preceding chapters. When we move from "heart" to "hand," we come upon several concepts—"rituals," "technology," and "addiction," which will be discussed in chapter 5. My emphasis on the contrast between stratified and evolutionary personality structures is vital for being able to use the extraordinary language in Figure 4-1 to help one solve problems, yet I have not depicted this contrast for purposes of simplification. All of the chapters, however, are organized with the contrast between stratified and evolutionary structures in mind.

Concerning the organization of chapter 4 in particular, let us recall the discussion—in the "Theoretical Framework" section of the introduction—of the importance of all three potentials of language—dichotomous, gradational, and figurative—for helping us to solve problems. Our focus in the first section of this chapter ("Stratified or Bureaucratic Worldview") will be on dichotomy. The final section ("Evolutionary Worldview") will focus on gradation and figurative speech. Further, the first section will emphasize Western culture; the final section, Eastern culture. Just as our overall orientation is to learn how to use all three potentials of language, so it is to examine just how Western and Eastern cultures can interact usefully with one another. Eastern culture emphasizes gradational and figurative language more than Western culture, which emphasizes dichotomous language. Thus, interaction with these two cultures can yield a broader approach to language in both of them.

Stratified or Bureaucratic Worldview

> "The wise man sees ahead. The fool walks in the dark." No doubt! But I know too, that one fate awaits them both. "The fool's fate," I thought to myself, "will be my fate too. Of what use my wisdom, then? This, too," I thought, "is vanity." Since there is no lasting memory for wise man or for fool, and in the days to come both will be forgotten; wise man, alas, no less than fool must die. Life I have come to hate, for what is done under the sun disgusts me, since all is vanity and chasing of the wind....
>
> What is, already was; what is to be, has been already.... Indeed, the fate of man and beast is identical; one dies, the other too, and both have the selfsame breath; man has no advantage over the beast, for all is vanity. Both go to the same place; both originate from the dust and to the dust both return. Who knows if the spirit of man mounts upward or if the spirit of the beast goes down to the earth? I have put all this to the test by wisdom, claiming to be wise; but wisdom has been beyond my reach. Reality lies beyond my grasp; and deep, so deep, who can discover it?...
>
> I come again to contemplate all the oppression that is committed under the sun. Take for instance the tears of the oppressed, with no one to protect them; the power their oppressors wield. No one to protect them!

> So, rather than the living who still have lives to live, I salute the dead who have already met death; happier than both of these is he who is yet unborn and has not seen the evil things that are done under the sun. I see that all effort and all achievement spring from men's mutual jealousy. This, too, is vanity and chasing of the wind....
>
> I find woman more bitter than death; she is a snare, her heart a net, her arms are chains; He who is pleasing to God eludes her, but the sinner is her captive.... I see this too under the sun; the race does not go to the swift, nor the battle to the strong; there is no bread for the wise, wealth for the intelligent, nor favor for the learned; all are subject to time and mischance. Man does not know his hour; like fish caught in the treacherous net, like birds taken in the snare, so is man overtaken by misfortune suddenly falling on him.... Vanity of vanities ... All is vanity (Jones, 1966: 856-857, 860-861, 864; Ecclesiastes 2: 14-18; 3: 15, 19-21; 7: 23-27; 8: 11-12; 12: 8).

This quotation is taken from Ecclesiastes—or "the Preacher"—one of the five "wisdom books" of the Old Testament. Rather than centering on religious beliefs, it emphasizes the broad wisdom derived from experience and attempts to offer a recipe for successful living. Thus, it illustrates Kelly's view of the individual as a kind of scientist, seeking to understand and control events. These passages were probably written by several individuals during the third century B.C.

The linguistic emphasis of this quote, and of Ecclesiastes as a whole, is on a dichotomous orientation to phenomena. For example, we have contrasts between the "wise man" and the "fool," between seeing ahead and walking in the dark, between "man" and "beast," between one's spirit moving upward and moving downward after death, between wisdom and an inability to understand reality, between the living and the dead, and between women and men. Such a one-sided emphasis of usage has its advantages, granting that its narrowness illustrates a stratified worldview. The authors are able to achieve a good deal of power in their efforts to communicate. We might recall once again the lines from Yeats's "The Second Coming," quoted in chapter 2:

> The best lack all conviction, while the worst
> Are full of passionate intensity.

We might see "the best" as those with an orientation to knowledge that is broad enough to include other potentials of language in addition to its dichotomies, namely, its potentials for gradational thought as well as for figurative thought. "The worst," given the narrowness of their focus on dichotomous thought, are in a good position to achieve "passionate intensity": they need not worry about shades of gray or about the limitations of their argument.

There are also other indications of a stratified worldview. We might note, for example, the view of women as "more bitter than death: she is a snare, her heart a net, her arms are chains." Another indication of such a worldview is the frequent statement that "all is vanity," implying a focus on people's feelings of self-importance or superiority in relation to others. This focus on hierarchy also becomes apparent in the writer's analysis of achievement: "I see that all effort and all achievement spring from men's mutual jealousy." "Jealousy" is a word that invokes a pattern of hierarchy, yet another indication of a stratified worldview. A further indication of a stratified worldview is the passage's orientation to change: "What is, already was; what is to be, has been already." There is, then, little importance attached to the time dimension, for change will not occur, and the individual is helpless in relation to any efforts to achieve change. A lack of interest in the time dimension—or Timeland, following the figure of speech presented in chapter 1—is also indicated by a focus only on what happens in the long run, not on what happens from one moment to the next: "wise man, alas, no less than fool must die."

In addition to the orientation of Ecclesiastes to a stratified or bureaucratic worldview, there is also the suggestion of a large gap between aspirations and their fulfillment. For example, much in the quote illustrates the centrality of achieving wisdom or learning. Yet the authors claim that "wisdom has been beyond my reach. Reality lies beyond my grasp; and deep, so deep, who can discover it?" Also, a lack of justice widely exists, implying once again an aspirations-fulfillment gap: "the race does not go to the swift, nor the battle to the strong; there is no bread for the wise, wealth for the intelligent, nor favor for the learned." This aspirations-fulfillment gap is itself associated with a stratified worldview, as we have seen illustrated in the foregoing chapters. For that worldview is sufficiently narrow to yield a divide between what people want and what they in fact are able to obtain.

By contrast, an evolutionary worldview is oriented to narrowing the aspirations-fulfillment gap in the short run by lowering short-term expectations, with a result that both aspirations and their fulfillment can be raised—one step at a time—in the long run. By so doing, we need not sacrifice fundamental long-term values that contribute to our large aspirations-fulfillment gap but, rather, can gradually work toward fulfilling them. For example, the passages in Ecclesiastes put forward aspirations for wisdom, for the protection of the oppressed, for an end to a focus on vanity with its elitist orientation to social stratification, and for "bread for the wise, wealth for the intelligent" and "favor for the learned" (implying higher pay for teachers and researchers, among others). Yet to move toward fulfilling such values, it is essential that we change nothing less than our stratified worldview.

Let us, then, continue our efforts to understand more fully the nature of our stratified worldview in this section. In the next section our focus will shift to how we can move to an alternative such as the evolutionary worldview. Jack Levin's doctoral dissertation in sociology at Boston University, "The Influence of Social Frame of Reference for Goal Fulfillment on Social

Aggression," was not developed to dip into the nature of a stratified and an evolutionary worldview, as we might think from its title. Nevertheless, it succeeded in doing so (Levin, 1968; see also Phillips, 1979: 185–187; 2002: 199–226). A title that conveys how it bears on worldviews might be "The Influence of Worldview on Prejudice."

Levin's procedures were broad. For example, he made much of the idea of "relative deprivation," which is the individual's awareness of a gap between certain aspirations and their fulfillment:

> The appearance of relative evaluation among Negro Americans may be having profound effects on the character of the civil rights movement in the United States and on the expectations which Negro Americans have about it. Even though the actual gains of Negroes have been occurring faster since 1940 than during any other period of Negro American history, the position of Negro Americans assumes a desperately deprived character when contrasted with the typical conditions of contemporary white Americans.... The result has been a keen sense of relative deprivation. (1968: 26)

This concept of relative deprivation overlaps with our own emphasis on an individual's growing aspirations-fulfillment gap, as initially presented in the introduction. By contrast, relative deprivation (1) centers on only some aspirations versus all aspirations, as developed in Figure I-1 and the foregoing chapters, and (2) is a phenomenon requiring the consciousness or awareness of the individual. Instead of burying such awareness, as illustrated by the Springdalers and as discussed in chapter 1, individuals who are experiencing relative deprivation remain aware of their aspirations-fulfillment gap. That in turn works to motivate them to take actions to narrow that gap. Here, then, is an immediate lesson for how an individual can proceed to change his or her worldview: become aware of how that worldview fosters one's aspirations-fulfillment gap. In other words, the individual can learn to experience feelings of relative deprivation when comparing—for example—his or her aspiration-fulfillment gap linked to a stratified worldview with the narrowing gap linked to an evolutionary worldview.

Levin accepted the importance of relative deprivation and the feelings of frustration generally accompanying it. He also opened up to greater complexity in the genesis of prejudice. He proceeded to contrast what he called two distinct "social frames of reference for goal fulfillment" but what we might call two worldviews corresponding to a stratified worldview and an evolutionary worldview:

> These alternative frames of reference for goal fulfillment can be identified as relative evaluation whereby the individual judges his personal performances relative to the productivity of other persons and groups, and self evaluation whereby the individual relies upon his other personal performances, past or present, as a standard of comparison. (1968: 20)

Levin thus saw prejudice not only as linked to feelings of relative deprivation but also linked to the individual's worldview. In particular, these two alternative frames of reference or worldviews correspond to the contrast that I have developed in the introduction and in the preceding chapters. Levin decided that an experiment coupled with a questionnaire would be the best way to address the question of how to test his ideas. He would begin with an initial measurement of degree of prejudice by using a questionnaire. Later he could launch his experiment, which would assess whether the individuals involved were oriented to a stratified or evolutionary worldview. It would also differentiate among groups of individuals: some would experience both frustration and relative deprivation, and others would not. He would also include a final measurement of degree of prejudice to assess the impact that worldview, frustration, and relative deprivation had on the individual's degree of prejudice. For example, would it take all three factors—a stratified worldview, frustration, and relative deprivation—to yield increased prejudice? Would those individuals with an evolutionary worldview prove to be resistant to such increased prejudice despite their experiencing frustration and relative deprivation?

Levin's questionnaire and experiment involved 180 freshmen and sophomores enrolled in two sections of introductory sociology at Boston University during the fall semester of 1965. His initial questionnaire—administered long before the experiment took place—measured attitudes toward Puerto Ricans using Osgood's semantic differential (1967):

> Directions: Place an "X" in one position between the adjectives of each scale (e.g., ___ : ___ : ___) to indicate how well these adjectives apply in general to Puerto Ricans. Your evaluation should reflect what you believe many of the members of this particular group tend to be (what the average Puerto Rican is like), and not necessarily what 100% of them are.

PUERTO RICANS

reputable	___ : ___ : ___ : ___ : ___ : ___ : ___	disreputable
knowledgeable	___ : ___ : ___ : ___ : ___ : ___ : ___	ignorant
intelligent	___ : ___ : ___ : ___ : ___ : ___ : ___	stupid
industrious	___ : ___ : ___ : ___ : ___ : ___ : ___	lazy
kind	___ : ___ : ___ : ___ : ___ : ___ : ___	cruel
clean	___ : ___ : ___ : ___ : ___ : ___ : ___	dirty
straightforward	___ : ___ : ___ : ___ : ___ : ___ : ___	sly
reliable	___ : ___ : ___ : ___ : ___ : ___ : ___	unreliable

Much later, when Levin began his experiment, he administered a questionnaire within the experiment with several series of paragraphs like the one below. Those checking 1 or 2, comparing themselves with others, were seen as illustrating relative evaluation or a stratified worldview. Those checking 3 or 4, comparing themselves with their own previous performances, were seen as illustrating self-evaluation or an evolutionary worldview:

> Mary was in her freshman year of high school and had just received a B on her first algebra examination. Joan, who was Mary's best friend, got an A on the same examination. The class average for the exam was C. Last year in junior high school, Mary had received a C in mathematics. This year, Mary's highest exam grade was an A in French.
>
> Without referring back to the paragraph on the preceding page [where the previous paragraph was located] what do you think is the most accurate way to describe Mary's grade on her first algebra examination? (check only one)
>
> 1___ Mary's algebra grade was higher than the class average
> 2___ Mary's algebra grade was lower than her best friend's grade
> 3___ Mary's algebra grade was higher than her last year's grade in mathematics
> 4___ Mary's algebra grade was lower than her grade in French

Frustration was introduced within the experiment by leading students to believe that they were taking an aptitude test for graduate school and then making it impossible for them to pass the test. Feelings of relative deprivation were influenced by giving students this note in their test booklets: "In similar groups of undergraduates at Boston College and Syracuse University, the average student was able to correctly complete 143 of the 150 items." Given the fact that no student completed more than 100 items, they were all led to believe that they did substantially worse than the average student at Boston College and Syracuse University.

The results of the Levin experiment strongly support the importance of the individual's worldview as a factor in the genesis of prejudice. All by itself, frustration did not yield increased prejudice against Puerto Ricans. And all by themselves, feelings of relative deprivation were not linked to increased prejudice. When these two factors were combined with a stratified worldview, however, the result was increased prejudice. That worldview influenced them to behave much like being on a seesaw relative to Puerto Ricans. Levin hypothesized that by putting down Puerto Ricans their own self-esteem would rise, if only temporarily, following their frustrating experience. The self-evaluators with at least the beginnings of an evolutionary worldview, by contrast, would have little to gain by putting down—in seesaw fashion—members of a minority group and attempting to raise themselves up as a result. For they would see themselves as being located on their own stairways rather than on seesaws, looking to their own previous or contemporary performances in an effort to improve on them.

Recognizing in advance the potential importance of his finding that an individual's worldview affects prejudice, Levin explored its implications with the aid of his questionnaire. He found, for example, that those with a stratified or bureaucratic worldview emphasized the importance of obedience to and respect for authority, and also the importance of conventionalism or conformity by contrast with any kind of deviant behavior. He also found that they tended to downgrade the capabilities of the human being, including pessimism about the possibilities of science for yielding understanding of human behavior. Overall, the Levin experiment gives us a clearer idea of the nature of a stratified or bureaucratic worldview as well as an evolutionary worldview. It also gives us some evidence as to the negative impact of a stratified worldview—by contrast with an evolutionary worldview—on prejudice. And it suggests some specific aspects of a stratified worldview, such as stratification, conformity, and opposition to the scientific method. Further, it illustrates ways in which those worldviews might be measured.

Evolutionary Worldview

> The mystery of language was revealed to me. I knew then that "w-a-t-e-r" meant the wonderful cool something that was flowing over my hand. That living word awakened my soul, gave it light, joy, set it free!
>
> —*Helen Keller,* The Story of My Life, *1902, ch. 4*

I believe that neither those students of Levin who resisted prejudice against Puerto Ricans despite their frustrations nor any other group has developed an evolutionary worldview. Yet even the beginnings of such a worldview—as demonstrated by the Levin experiment—can have enormous repercussions. And if we open our eyes to the breadth of human experiences, we can detect such beginnings in many places. For example, the experience of the blind and deaf Helen Keller upon beginning to learn the nature of language—as just quoted—suggests the importance of language for our learning to solve the fundamental problems that threaten contemporary society. Keller also illustrates the potential that all of us humans have, given the incredible problems that she somehow was able to overcome.

If Ecclesiastes and the Levin experiment illustrate Western thought, we might do well to turn at this point to Eastern thought to develop the breadth of perspective that we require in confronting a problem so vast as the nature of our worldview along with an alternative worldview. The use of pictographic written language in China and Japan, with its derivation from pictures of objects by contrast with the abstract alphabet used throughout Western society, suggests substantial differences between Eastern and Western thought. This is exactly what I am looking for. Pictographs suggest more of an emphasis on figures of speech within China and Japan than what is found in Western society, even assuming that contemporary individuals no longer pay attention to the original pictures on which their written language was based. Western societies—following the approach within Ecclesiastes—appear to have emphasized dichotomies in ordinary language. By exploring Eastern thought, then, we can extend that orientation to include figures of speech. This is also suggested by the overall difference between Western and Eastern philosophy, for it is the latter rather than the former that emphasizes myths, stories, and sayings, or figures of speech. This contrasts with the greater attention to abstract ideas and logical relationships among concepts in Western philosophy.

Given my interest in examining the nature of an evolutionary worldview, I turn to the idea of karma, which is important in both classical Indian or Hindu philosophy as well as Buddhist philosophy. For some background, we should understand the Indian assumption that all organisms are reincarnated after death. In other words, organisms lead a succession of lives. Abraham Kaplan, an American philosopher who combined an interest in the Western philosophy of pragmatism with interest in Eastern philosophies and religions, writes about karma as follows:

> What *is* essential [within classical Indian philosophy] is that succession of lives—and more to the point, the succession of experiences within any of them—is not random, but proceeds in accord with an inflexible law known in Indian thought as the law of *karma.* It can best be understood, I think, as a principle of moral causation. Western philosophy has relied heavily on a concept of causality as a principle of natural necessity: everything happens just as and when it does because of the workings of causes which are not to be denied their effects. Karma is, as it were, an application of this idea to the sphere of conduct, and especially with regard to the moral quality of the acts performed.... We may say that karma is the collective force of a man's past actions.... In short, the law of karma has roughly the same content as the Greek aphorism that man's character is his fate. What happens to a man is an inevitable outcome of what he is, and this in turn is a necessary consequence of his past actions....
>
> The problem of life, then, for the Indian philosophers, is to work out your karma to its end. For it has an end, a final release from *samsara* [the endless round of existence from birth to death to rebirth]. This final release the Indians call *moksha,* liberation from the body of death. It is better translated as "emancipation" rather than "salvation." For man is not infected with original sin, innate depravity, or anything of that kind. We do not face a damnation from which we need to be saved....
>
> Yoga conceives of the body as the instrument or expression of the mind and soul: it is not a prison for them, not an obstacle to their free working. Yoga aims at achieving a fullness of expression, a mastery of the instrument which is the body.... In these terms, Indian thought distinguishes three yogas, which have their counterparts in all major world religions and moral philosophies. The first of these is *bhakti-yoga,* the path of love, devotion, or faith ... bhakti-yoga is accessible to everyone, whatever his talents or station in life.... Among institutionalized religions, bhakti-yoga is emphasized, for instance, by the Hasidim in Judaism, the Sufi in Islam,

> Zen in Buddhism, and the Pietists in Christianity. Second is the path of *karma-yoga,* the way of moral action. Here we may think of Gandhi, and in the West of Lincoln and Schweitzer.... This is what we know in the West as the path of works.... Most characteristically Indian is the third yoga, *jnana-yoga,* the path of understanding.... This is the emphasis in *Ecclesiastes* and various of the Apocrypha and gnostic writings, and in such philosophers as Plotinus and Philo Judaeus. (Kaplan, 1961: 212-214, 224-226)

The concept of karma within Indian philosophy is a fundamental contribution to the idea of an evolutionary worldview if we divorce it from nonscientific ideas and link it to scientific ideas developed within Western society. It was Rudyard Kipling who suggested an East-West synthesis of ideas:

> Oh, East is East, and West is West, and never the twain shall meet,
> Till Earth and Sky stand presently at God's great Judgement Seat;
> But there is neither East nor West, Border, nor Breed, nor Birth,
> When two strong men stand face to face, tho' they come from the ends of earth!
>
> —*Rudyard Kipling,* "The Ballad of East and West," *1892*

Kipling's idea that "East is East, and West is West, and never the twain shall meet" conveys the reality of the differences between Eastern and Western thought. If indeed Easterners and Westerners can find a way to become "two strong men," then they can learn to integrate their ideas, and that synthesis can prove to be most important for their learning to relate to one another.

"Karma" emphasizes a gradational approach to language, by contrast with the dichotomous emphasis in Ecclesiastes. When we link these two approaches, the resulting twofold approach to language makes more use of its potential. Further, Indian philosophy relies far more on figures of speech than does contemporary Western philosophy, pointing us toward a third potential of language. Once again, the union of that potential with the other two opens us up to language's full potential. There is also much more to the idea of karma, however, that can contribute to an evolutionary worldview. Western physical science emphasizes causality within physical phenomena, and there is very little attention to social science within Western culture. But the idea of karma brings causality squarely into human phenomena and suggests the centrality of social science for contemporary culture.

If we now turn to the last paragraph of the previously quoted analysis of classical Indian philosophy centering on yoga, we find that the different species of yoga parallel our own distinctions among "heart," "hand," and "head." For bhakti-yoga ("the path of love, devotion, or faith") suggests the "heart," karma-yoga ("the way of moral action") suggests the "hand," and jnana-yoga ("the path of understanding") suggests the "head." Yet remaining to develop is a metaphysical stance that gives weight to all three, a stance that I am pointing toward with the idea of an evolutionary worldview.

Classical Indian philosophy was followed by the development in India of another world religion, Buddhism, which also emphasizes the importance of karma. Kaplan states that the Buddhist approach to karma is more comprehensive—and thus more scientific—than the approach to karma within classical Indian philosophy:

> Buddhism takes karma even more seriously, if possible, than the orthodox Indian thinkers. For it generalizes the law of karma into a principle of universal causality. We may note that such a conception was not very important in Western thought till, say, the end of the eighteenth century; and even then, causality was thought not to apply strictly in the biological sphere, or, at the very least, in the sphere of human behavior. But for the Buddha, the law of causality is the second of the Four Noble Truths.... Everything happens as it does in accord with a causal principle. If there is dukkha in human life, something must cause it, and if this cause were removed its effect also would disappear.... The world process is a network of causal relationships.... Buddhism turns on the grasp of causal relations, initially, in the world process, and ultimately, in the life of man, ... if ... life is dukkha, there is a cause which has brought about this effect. The solution to the problem of life lies in identifying and removing this cause. (Kaplan, 1961: 248–250)

This focus of Buddhism on a principle of universal causality, and not just on the experiences of the individual in his or her journey through life, opens up to the full breadth of the Web and Part/Whole Approach to the scientific method. There are other problems in the world than that of how the individual can evolve. A great many of them have to do with the power of social structures, granting that individual structures remain a central force in solving those problems. By opening up to universal causality, Buddhism becomes very modern in its commitment to a broad scientific method.

Just as we have dipped into classical Indian philosophy and Buddhism to gain further understanding of a stratified and an evolutionary worldview, so can we continue to gain such understanding by dipping into works from Western authors. We might start with Ervin Laszlo, a philosopher who wrote *Introduction to Systems Philosophy: Toward a New Paradigm of Contemporary Thought* (1973). We might see him as emphasizing the importance of learning about our metaphysical stance or worldview as well as providing an alternative to it:

> We are at a cross-roads.... And fight we must. Not one another, but *for* one another. Not with violence, but with perseverance and courage. The fight is not for national victory, of one people over another, but for the

> persistence of the kind of life which we call human, and which incorporates our finest traditions and achievements. This is not a fight *against*, but a fight *for*. And it can be lost just as readily as any fight against an enemy. Defeat here will be marked by the extinction of our species or, at best, by the downfall of human civilization and its replacement by a form of existence at which we can only shudder. It may be the anti-society of maximum regimentation described by Orwell's *1984*, or an uncoordinated aimless physical survival in a state of near chaos. Victory will signify the continuation of civilization as we know it, with material and spiritual resources available to all....
>
> Metaphysics is perhaps the oldest of human cognitive endeavors: It shades imperceptibly into myth in antiquity, and into philosophy and science in modern times. Metaphysics, as religion, gives answers to questions concerning the ultimate nature of reality. It differs from religion in that its propositions rest not on faith.... The contribution of philosophical analysis ... becomes real only when it is placed within the context of our scientific (i.e., imaginatively postulated and rigorously tested) knowledge of natural events.... The most consistent as well as most general paradigm available today to the inquiring, ordering mind is the systems paradigm. Explicated as a *general theory of systems,* both "physical" and "mental," and applied to the analysis of human experience and its problems, it constitutes *systems philosophy.* This, then, is the conceptual framework within which classical philosophical analyses of experience, knowledge, art, religion, valuation, freedom, morality, dignity, and many other facets of philosophical interest, can take place. (Laszlo, 1973: 281–282, 291, 297–298)

In the first paragraph Laszlo writes about the existence of a crisis in 1973 society that can lead to "the extinction of our species or, at best, by the downfall of human civilization and its replacement by a form of existence at which we can only shudder." He wrote those words after the Cuban Missile Crisis and during the protracted cold war between the Soviet Union and the United States, a war that could have easily led to nuclear warfare. I believe that those threats to the contemporary world are even greater today, given the continuing development of deadlier weapons of mass destruction, the increasing possibilities for small groups dedicated to using them to obtain such weapons, and the continuing failure of social scientists to integrate their knowledge so as to provide political leaders and others with the basis for solving world problems.

In the second paragraph Laszlo writes of the importance of a new and very broad metaphysical "paradigm" or worldview, presumably to integrate available knowledge so as to gain understanding of existing threats to civilization and learn to counter them. That approach is much the same as my own, as developed in the foregoing materials. More specifically, Laszlo puts forward the idea of a "systems philosophy" as the basis for such a paradigm or worldview. It was in 1954 that Norbert Wiener, a physicist, published *The Human Use of Human Beings: Cybernetics and Society.* In 1967 sociologist Walter Buckley published *Sociology and Modern Systems Theory.* This was followed in 1968 by the biologist Ludwig von Bertalanffy's *General System Theory.*

Just as Laszlo's concern with the future of society is very well taken, so is his conviction as to the importance of developing a new metaphysical stance—closely allied with scientific research—that is broad enough to confront human complexity and threatening problems in contemporary society. Further, systems theory—as illustrated by Buckley's book, which is discussed at length in *The Invisible Crisis of Contemporary Society* (Phillips and Johnston, 2007)—contributes substantially to our understanding of the complexity of human phenomena. And by so doing, it contributes to our understanding of what is involved within an evolutionary worldview. Yet there are serious problems with the development of a metaphysical stance based on systems theory. For that theory is linked far more closely to physical and biological science than to social science. This is illustrated by Laszlo's failure to even cite sociology's major classical figures, such as Marx, Durkheim, Weber, and Simmel. Further, his orientation to the history of philosophy appears to be one-sided, given his omission of references to such figures as the pragmatist philosophers (Peirce, James, Dewey, and Mead) as well as many major philosophers like Plato, Aristotle, Descartes, Kant, Hume, Hegel, Nietzsche, and Wittgenstein. Further, Laszlo has little to say about the role of language in developing a new worldview.

A philosopher who contributes substantially to an understanding of an evolutionary worldview is Charles Sanders Peirce, the founder of the philosophy of pragmatism (also discussed in Phillips and Johnston, 2007). Peirce began writing in 1877, continuing until just before his death in 1914. Like Thomas Kuhn—discussed in the introduction—he was a student of both physical science and history, especially the history of philosophy. His immersion in physical science was deep, including 40 years of work in physics, chemistry, and mathematics as well as contributions to those fields.

Peirce's essay, "The Fixation of Belief" (1877/1955: 5-22), conveys an idea that is central to his—and my own—understanding of the scientific method:

> The irritation of doubt is the only immediate motive for the struggle to attain belief.... With the doubt, therefore, the struggle begins, and with the cessation of doubt it ends. Hence, the sole object of inquiry is the settlement of opinion ... as soon as a firm belief is reached we are entirely satisfied, whether the belief be true or false.... That the settlement of opinion is the sole end of inquiry is a very important proposition. It sweeps away, at once, various vague and erroneous concepts of proof.... Some philosophers have imagined that to start an inquiry it was only necessary to utter a question whether orally or by setting it down upon paper,

> and have even recommended us to begin our studies with questioning everything! But the mere putting of a proposition into the interrogative form does not stimulate the mind to any struggle after belief. There must be a real and living doubt, and without this all discussion is idle.... When doubt ceases, mental action on the subject comes to an end; and, if it did go on, it would be without purpose. (1877/1955: 10–11)

Here is the first and most important step of the scientific method: awareness of and definition of a problem. This understanding of the scientific method suggests an evolutionary perspective: we can continue to learn without any limit, for after we attain partial understanding of a phenomenon, we come to doubt the completeness of that understanding, and so on indefinitely. Peirce ranges very widely in his effort to clarify the nature of ideas by bringing in the phenomenon of perception, as discussed in chapter 2:

> Our idea of anything is our idea of its sensible effects; and if we fancy that we have any other we deceive ourselves.... It is absurd to say that thought has any meaning unrelated to its only function.... It appears, then, that the rule for attaining ... clearness of apprehension is as follows: Consider what effects, that might conceivably have practical bearings, we conceive the object of our conception to have. Then, our conception of these effects is the whole of our conception of the object. (1878/1955: 31)

"Sensible effects" have to do with what we perceive, and our perception is involved in everything that we think, feel, and do. Here Peirce alerts us to understand thought not only in relation to feelings and action but also with respect to the perceptions that underlie "head," "heart," and "hand." As taken up in chapter 2, a stratified worldview—which is a system of beliefs—implies outward perception, whereas an evolutionary worldview implies inward-outward perception.

Peirce carries further his view of the scientific method as a process without any end. In so doing he comes up with perhaps his best known statement:

> Upon this first, and in one sense this sole, rule of reason, that in order to learn you must desire to learn, and in so desiring not be satisfied with what you already incline to think, there follows one corollary which itself deserves to be inscribed upon every wall of the city of philosophy: Do not block the way of inquiry. (1896/1955: 54)

For Peirce, then, inquiry should go on indefinitely. And, I might add, so should the overall development or evolution of the individual. A fundamental problem getting in the way of an evolutionary worldview is a metaphysical stance or worldview. Peirce writes:

> There is and can be no doubt that this immature condition of Metaphysics has very greatly hampered the progress of ... psychology ... linguistics, anthropology, social science, etc. To my mind it is equally clear that defective and bad metaphysics has been almost as injurious to the physical sciences.... The common opinion has been that Metaphysics is backward because it is intrinsically beyond the reach of human cognition.... But Metaphysics, even bad Metaphysics really rests on observations, whether consciously or not; and the only reason that this is not universally recognized is that it rests upon kinds of phenomena with which every man's experience is so saturated that he usually pays no particular attention to them. (1898/1955: 310–311)

Peirce's argument that lack of progress in social science is largely based on what he calls the "immature condition of Metaphysics" meshes with my own argument that a stratified metaphysical stance or worldview impedes our ability to solve problems, and that this can be reversed by a shift to an evolutionary worldview.

We have yet to emphasize the "hand" component of personality structure, that is, the individual's pattern of actions and interactions. That is our task in chapter 5. More specifically, Figure 4-1 includes under "hand" "rituals," "technology," and "addiction." How can these ideas help us to understand problems within personality structures and directions for their solution at this time in history? Chapter 5's title is "'Hand': Addiction Versus Meaningful and Expressive Rituals." How is addiction related to a stratified worldview, and how are meaningful and expressive rituals linked to an evolutionary worldview? As yet we have only taken a few steps toward understanding the nature of these two metaphysical stances, let alone how we might proceed to move from the former to the latter.

Chapter 5

"Hand"

Addiction Versus Meaningful and Expressive Rituals

Marx's eleventh thesis on Feuerbach, written in 1845, reads: "The philosophers have only interpreted the world in various ways; the point, however, is to change it" (Marx, 1845/1972: 109). In this "hand" chapter I center on changing structures and, as a result, changing the world. Yet to achieve such change or an emphasis on "hand," it is crucial to change both "heart" and "head," which has been the focus of chapters 3 and 4, respectively. I might as a result alter Marx's eleventh thesis on Feuerbach to read: "The philosophers and social scientists have only interpreted the world in various ways; the point, however, is to interpret it, to commit to that interpretation, and to change it." Effective action requires understanding and motivation. The philosophers of Marx's day have been joined by the social scientists in our own day, but they have developed largely unrelated pieces of knowledge, failing to yield integrated understanding of our problems. That integration, and the commitment that can foster it, are crucial to follow through on Marx's thesis.

Figure 4-1 included the concept "rituals" as illustrating "hand," by contrast with "heart" and "head." In the glossary a "ritual" is a "persisting pattern of action that is meaningful and expressive to some degree." Two types of rituals that we shall emphasize, listed in Figure 4-1, are "technology" and "addiction." "Technology" is "a persisting pattern of action for solving problems" and is closely linked to our understanding of the scientific method. Long before the scientific method was invented—a method that emphasizes building on recorded knowledge—people used technologies, such as the wheel and the plough, to solve their everyday problems. And we all continually make use of technologies as we go about our ordinary business with little regard to recorded knowledge in our libraries. But since the sixteenth century more and more attention has been paid to what others have learned in the past as a result of the invention of writing and, in particular, of the printing press in the middle of the fifteenth century. As a result, we tend to think of "technology" as exclusively "scientific technology." Yet we would do well to understand "technology" as a much broader category, granted the enormous importance of "scientific technology" over these past centuries. As for "addiction," its relationship to rituals will be emphasized in this introduction.

The idea of developing meaningful and expressive rituals by contrast with patterns of addictive behavior can be illustrated by a study of concentration camps in Germany, the Soviet Union, the United States (the internment of Japanese civilians), and Japan between 1935 and 1955 (Knottnerus, 2002). Knottnerus quotes Janusz Bardach, who survived the Soviet labor camps and later described his experiences in *Man Is Wolf to Man: Surviving the Gulag*:

> At four A.M. the ringing rail sounded for us to get up. Despite my fatigue and the cold, I kept the exercise routine I had followed at home and in the Red Army, washing my face and hands at the hand pump. I wanted to retain as much pride in myself as I could, separate myself from the many prisoners I had seen give up day by day. They'd stop caring first about their hygiene or appearance, then about their fellow prisoners, and finally about their own lives. If I had control over nothing else, I had control over this ritual, which I believed would keep me from degradation and certain death. (Bardach, 1998: 130)

Rituals are not restricted to religious rituals: they can be any persisting or repetitive pattern of action or interaction. In this way we define rituals to include almost all of our everyday behavior, which we would ordinarily call our habits of action, as distinct from our habits of thought (for example, beliefs and assumptions) or habits of feeling (for example, values). Following our definition of a ritual as a persisting pattern of action that is meaningful and expressive to some degree, rituals can vary greatly in the degree of meaningfulness and expressiveness. These are structures that are part of our personality structure, just as are our beliefs, assumptions, and values. It is our structures that are so crucial in shaping our moment-to-moment behavior in particular situations. By learning to employ these terms—"rituals," "beliefs," "assumptions," and "values"—to invoke our habitual behavior in everyday life, we also learn to organize our thinking about the behavior of the individual in a systematic way. Let us recall Figure 4-1, where personality structure was depicted as tying together values, beliefs and assumptions, and rituals. When we examine social structures in chapters 6–8, we will learn to extend this systematic approach to the behavior of groups.

To return to Bardach's experiences in the Gulag, he distinguished between two kinds of routines or rituals: "washing my face and hands at the hand pump," and the routines of many other prisoners: "They'd stop caring first about their hygiene or appearance, then about their fellow prisoners, and finally about their own lives." They had given up on the rituals that had been meaningful throughout their lives, rituals that also had been satisfying emotionally, and restricted themselves to the meaningless rituals of what the Gulag forced them to do. By contrast, Bardach defined his ritual of washing his hands and face at the hand pump as both meaningful and expressive for him. It gave him a sense of "control" over his situation. It may seem to us like a very small thing, but he sees it as having been most important for his survival. Thus, these two kinds of rituals varied greatly in their degree of meaningfulness and expressiveness.

We can gain further insight into these two kinds of rituals from Gustav Herling, also interned in a Soviet concentration camp. Herling contrasts his own behavior, such as mending clothes, writing letters, or visiting friends in the barracks in the evening, with the behavior of the majority of prisoners:

> All these [his own] activities had one quality in common—they imitated the normal occupations of a free life. Our behavior was a parody of the gestures, habits and responses of our former existence, observing the symbolic ritual of a dimly remembered routine.... But the majority ... left their bunks during the evening only to fill up the hollowness of their stomachs with a pint or two of the inevitable hot water.... The majority of prisoners ... deluded themselves that this suicidal form of relaxation strengthened the organism of their bodies. In the camp the normal process was reversed: inertia and apathy hastened death, while any form of activity postponed it for an unforeseen period. (Herling, 1951: 143-144; see also Knottnerus, 2002)

Although we might see almost all of our behavior as meaningful and expressive at least to some degree, some behavior covers far more of the ground of our personality structure than other behavior. And thus we may see it as more meaningful and expressive, at least with respect to the personality structure as a whole. Herling's activities in the evening had to do with a wide range of his interests. In his words, "they imitated the normal occupations of a free life." By contrast, the ritualistic or routine behavior of the majority of prisoners was extremely limited with reference to the range of their former interests: "the majority ... left their bunks during the evening only to fill up the hollowness of their stomachs with a pint or two of the inevitable hot water." There is a term that we might employ for such ritualistic behavior that is far less meaningful and expressive for an individual's personality as a whole than what Herling did: "addiction" or addictive behavior. We might, then, define "addiction" as an individual's subordination of individuality to dependence on external phenomena. We may see addictive behavior, then, as far less meaningful and expressive than nonaddictive behavior.

"Addiction" may involve a physiological dependence on some substance such as drugs, alcohol, or cigarettes. We can also include psychological dependence, however, such as dependence on work ("workaholism"), watching television, working out, sexual experiences, overeating, avoiding eating insofar as possible, or just about any other activity that dominates much of one's way of life and thus narrows one's experiences. Most of those experiences can be valuable in moderation, but it is an overriding dominance of some external dependence that narrows the range of one's possibilities to which "addiction" refers. We might have reference here to the many different twelve-step groups—following the lead of Alcoholics Anonymous—that have been organized to help the individual cope with psychological addictions, illustrating the scope of this problem. For example, there is Batterers Anonymous, Debtors Anonymous, Depressives Anonymous, Emotions Anonymous, Families Anonymous, Gamblers Anonymous, Molesters Anonymous, Sex Addicts Anonymous, Workaholics Anonymous, Shoplifters Anonymous, Spenders Anonymous, and Women with Multiple Addictions. These, however, are no more than a small number of examples. My orientation to addiction is not that many of us are addicted to one thing or another. Rather, my orientation is similar to Horney's (1937) analysis of the sources of neurotic behavior in *The Neurotic Personality of Our Time.* It appears that modern society itself fosters an addictive personality and that we are all more or less addicted. For I see addiction as linked closely to our stratified worldview, which points us all outward versus both inward and outward.

As we proceed in this chapter and the ones to follow to open up to more ideas, it is essential that we return to the general framework sketched in Figure I-2, with its depiction of the scientific method linked to worldview, aspirations-fulfillment gap, and personal and social problems. Continuing to return to that framework will help us integrate new ideas with earlier ones, by contrast with the largely unconnected ideas throughout the academic world, as well as within each of us, as a result of our stratified worldview. This will be a habit of thought that can help guide us to develop habits of action or rituals in everyday life that—following Bardach's and Herling's experiences in the Gulag—can become meaningful and expressive.

Addiction

We can summarize the key elements of the previous analysis of the nature of addiction with a few quotes. Jean-Baptiste Moliere, the seventeenth-century French comic playwright, wrote that "one should eat to live, not live to eat," for if we "live to eat" we learn to ignore our possibilities for living a full life. Carl Gustav Jung, the twentieth-century Swiss psychologist, wrote: "Every form of addiction is bad, no matter whether the narcotic be alcohol or morphine or idealism," sensitizing us to an understanding of addiction as encompassing far more than substance abuse. Calvin Coolidge, president of the United States from 1923 to 1929, alerted us to the importance of actions that persist over time, which is fundamental to all rituals:

> Nothing in the world can take the place of persistence. Talent will not; nothing is more common than unsuccessful men with talent. Genius will not; unrewarded genius is almost a proverb. Education will not; the world is full of educated derelicts. Persistence and determination are omnipotent. The slogan "press on" has solved and always will solve the problems of the human race. (quoted in Knowles, 2004: 242)

Our focus in this section will be on the basis for the development of addictive behavior, our largely invisible aspirations-fulfillment gap, fostered by our stratified worldview. Instead of learning to face up to problems that we don't know how to solve, we run away from them by developing our many varieties of addiction. We begin with a selection from Hermann Hesse's 1943 novel *The Glass Bead Game (Magister Ludi)*, written during World War II. The book is an allegory of the "game" played by the academic world. Professors do succeed in keeping alive and extending the wisdom of all ages, which is exactly what Hesse had been doing during most of his life. Yet at the same time professors generally remain in their ivory towers, avoiding any involvement in the pressing problems of the day. They remain addicted to a focus on the "head" while largely ignoring the "heart" and the "hand." Hesse had a change of heart during that war. His hero in *Magister Ludi*—Joseph Knecht, master of the Glass Bead Game and leader of the intellectual order of Castalia—submits a letter of resignation to his ruling board:

> I have begun to doubt my ability to officiate satisfactorily because I consider the Glass Bead Game in a state of crisis.... Here I am sitting in the top story of our Castalian edifice ... and instinct tells me, my nose tells me, that down below something is burning, our whole structure is imperiled, and that my business now is not to analyze music or define rules of the Game, but to rush to where the smoke is....
>
> The history of societies shows a constant tendency toward the formation of a nobility as the apex and crown of any given society.... If, now, we regard our Order as a nobility and try to examine ourselves to see ... to what extent we have already been infected by the characteristic disease of nobility—*hubris,* conceit, class arrogance, self-righteousness, exploitativeness ... we may be seized by a good many doubts.... In brief, this Castalian culture of ours ... tends somewhat toward smugness and self-praise, toward the cultivation and elaboration of intellectual specialism.... Historically we are, I believe, ripe for dismantling. And there is no doubt that such will be our fate, not today or tomorrow, but the day after tomorrow.... Critical times are approaching; the omens can be sensed everywhere; the world is once again about to shift its center of gravity. Displacements of power are in the offing. They will not take place without war and violence....
>
> I herewith request the Board to relieve me of my office as Magister Ludi and entrust to me an ordinary school, large or small, outside in the

country; to let me staff it with a group of youthful members of our Order. (319, 321–322, 328–329, 335)

Hesse, experiencing the horrors of World War II, had come to see the work of academics largely as a way of escaping from—rather than confronting—growing problems throughout the world. And he saw social stratification, or the patterns of hierarchy prevalent throughout the academic world, as fundamental to this situation, illustrated by "*hubris,* conceit, class arrogance, self-righteousness, exploitativeness ... smugness and self-praise ... the cultivation and elaboration of intellectual specialism." Yet what we must learn to do is to understand such behavior and not simply condemn it. From the perspective of Figure I-2, the Glass Bead Game of the academic world is a focus on the "head" while largely avoiding the "heart" and the "hand." It is an addiction to avoiding action on pressing problems. While such behavior is temporarily satisfying to academics, in common with the temporary satisfactions associated with addictive behavior in general, world problems are continuing to increase. That is illustrated by a growing aspirations-fulfillment gap, and it appears to be just a matter of time before world chaos will take over our present way of life.

Yet all of that knowledge within the academic world, based on achievements of the human race over many centuries, need not continue to be shut up within the ivory tower of our libraries. Following World War II, Erich Fromm—a neo-Freudian psychoanalyst who shared Karen Horney's interest in culture—criticized people in general for running away from their faith in earlier ideals. Here, the addiction involved is equally an avoidance of the hand, or of not acting to address escalating problems:

> A spirit of pride and optimism has distinguished Western culture in the last few centuries: pride in reason as man's instrument for his understanding and mastery of nature; optimism in the fulfillment ... of the greatest happiness for the greatest number.... Yet modern man feels uneasy and more and more bewildered.... While becoming the master of nature, he has become the slave of the machine which his own hands have built. With all his knowledge about matter, he is ignorant with regard to the most important and fundamental questions of human existence: what man is, how he ought to live, and how the tremendous energies *within* man can be released and used productively. The contemporary human crisis has led to a retreat from the hopes and ideas of the Enlightenment under the auspices of which our political and economic progress had begun. The very idea of progress is called a childish illusion, and "realism," a new word for the utter lack of faith in man, is preached instead. The idea of the dignity and power of man, which gave man the strength and courage for the tremendous accomplishments of the last few centuries, is challenged by the suggestion that we have to revert to the acceptance of man's ultimate powerlessness and insignificance. This idea threatens to

> destroy the very roots from which our culture grew.... He [Man] reverts to a position which the Greek Enlightenment, Christianity, the Renaissance, and the eighteenth-century enlightenment had already overcome. The demands of the State, the enthusiasm for magic qualities of powerful leaders, powerful machines, and material success become the sources for his norms and value judgments.... Are we to leave it at that? (1947: 13-17, 251)

Jung had suggested that "every form of addiction is bad, no matter whether the narcotic be alcohol or morphine or idealism." We might also see one type of addiction as involving the avoidance of idealism and the substitution of a materialistic way of life, for that is what Fromm is writing about. It is also an avoidance of optimism about the possibilities of the human being, a hard-won optimism based on a long-term history that includes "the Greek Enlightenment, Christianity, the Renaissance, and the eighteenth-century enlightenment." Yet it was an optimism, associated with the "revolution of rising expectations"—or the upper curve of Figure I-1—that failed to take into account sufficiently the bottom curve, which departed more and more from the top curve to create an increasing aspirations-fulfillment gap. Neither did that optimism open up to the stratified metaphysical stance or worldview lying behind and shaping that increasing aspirations-fulfillment gap. Fromm's concern here extends beyond the academic world to include modern society in general. Although he retains his own optimistic stance about human possibilities, he sees a widespread growth of pessimism or, in the language of this chapter, "addiction" to a materialistic way of life as well as to social stratification, key elements of a stratified worldview: "The demands of the State, the enthusiasm for magic qualities of powerful leaders, powerful machines, and material success become the sources for ... norms and value judgments." Just as in the case of addictions in general, this kind of addiction closes down the individual's possibilities for living a full life. It yields a dependence on external phenomena that requires—just as in the case of the Springdalers—"falsification of memory" and "techniques of particularization."

As we continue to examine the forces producing addiction throughout contemporary society, let us jump over the remainder of the twentieth century to the end of 2006. On the cover of *Time* magazine's "Person-of-the-Year" issue is a computer screen covered by a piece of Mylar that reflects the reader's own image. *Time* sees the reader, especially everyone using a computer, as the person of the year, noting on the cover: "**You**. Yes, you. You control the Information Age. Welcome to your world." *Time*'s explanation for its choice suggests that "individuals are changing the nature of the information age, that the creators and consumers of user-generated content [such as what appears on YouTube] are transforming art and politics and commerce, that they are the engaged citizens of a new digital democracy ... [and] this new global nervous system is changing the way we perceive the world. And the consequences of it all are both hard to know and impossible to overestimate."

On the surface it would appear that *Time* is answering Erich Fromm. Supposedly, all is not lost. The individual's possibilities—along with the long and optimistic tradition of Western society—are not forgotten. Pessimism does not rule the world. The individual has access to the Internet and can freely express self and learn from everyone else. Yet is using the Internet little more at this time than just another way of escaping from world and personal problems, that is, just another addiction? John Berger, in his analysis of advertising on television in chapter 2, claimed that "publicity turns consumption into a substitute for democracy. The choice of what one eats (or wears or drives) takes the place of significant political choice. Publicity helps to mask and compensate for all that is undemocratic within society." Granting that the possibilities of the Internet for education as well as for helping us to make truly significant choices are great, is it in fact true that our use of the Internet at the present time contributes very little to either education or fundamental choices? Is it the case that the Internet—along with the mass media—generally follows Fromm's indictment of modern society in reinforcing "the demands of the State, the enthusiasm for magic qualities of powerful leaders, powerful machines, and material success . . ."?

Frank Rich, the columnist for *The New York Times* who wrote *The Greatest Story Ever Sold* (2006a)—discussed in chapter 2—sees *Time*'s conclusion that "you control the Information Age" as "a bountiful gift of mirth to America":

> As of Friday morning, "Britney Spears Nude on Beach" had been viewed 1,044,776 times by YouTube's visitors. The count for YouTube video clips tagged with "Iraq" was 22,783. Not that there is anything wrong with that. But *compulsive* blogging and free soft-porn are not, as *Time* would have it, indications of how much you, I and that glassy-eyed teenage boy hiding in his bedroom are in control of the Information Age. They are indicators instead of how eager we are to flee from brutal real-world information that makes us depressed and angry. This was the year Americans escaped as often as they could into their private pleasure-pods. (2006b: 8wk; italics added)

However, Rich does claim that after some reflection he believes that *Time* actually did make the right choice for the Person of the Year. The choice was made for the wrong reasons, however. Each of us is indeed the Person of the Year, although certainly not because we control the information age. For the year has been made up of a series of escapes from taking responsibility for confronting the fundamental problems of the information age. It is our addiction to avoiding those problems—while turning to "Britney Spears Nude on Beach."

Yet how might we learn to take such responsibility in 2007 and beyond? Instead of our addiction to escapes in 2006, what might we proceed to accomplish? More specifically, how might we develop rituals that are meaningful and expressive instead of the addictive rituals that were prevalent during 2006? In the next section we take up alternatives to our addictive behavior, taking into account the enormous difficulties that are involved. Our worldview and our aspirations-fulfillment gap are not unchangeable. We can indeed learn to alter them and, along with that, learn to confront the exceedingly complex problems of our times.

Meaningful and Expressive Rituals

> A man who reviews the old so as to find out the new is qualified to teach others.
>
> —*Confucius,* Analects, *ch. 2, v. 11*

> Is it not a pleasure to learn and to repeat or practice from time to time what has been learned? Is it not delightful to have friends coming from afar?
>
> —*Confucius,* Analects, *ch. 1, v. 1*

> At fifteen my mind was set on learning. At thirty my character had been formed. At fifty I knew the Mandate of Heaven. At sixty I was at ease with whatever I heard. At seventy I could follow my heart's desire without transgressing moral principles.
>
> *Confucius,* Analects, *ch. 2, v. 4*

Most of this book emphasizes the importance of structures, as illustrated by our focus on rituals in this chapter, on beliefs and assumptions in chapter 4, and on values in chapter 3. This emphasis implies the centrality of a long-term orientation to the dimension of time, or—figuratively, as developed in chapter 1—to Timeland. After all, we humans are the product of some fourteen billion years of evolution. Our focus on the increasing aspirations-fulfillment gap centers on no less than five centuries, as depicted in Figure I-1. And our emphasis on following the ideals of the scientific method throughout this book also requires that we build on whatever understanding of human behavior has been achieved in the past. This conservative interest in the past was developed in great detail by the Chinese philosopher and teacher Kung-Fu-tzu, whom we know as Confucius, who lived in the sixth and fifth centuries B.C. His ideas can help us to understand the nature of the rituals to be found in our everyday lives. Further, they can help us to learn how to move away from ritualistic behavior that is addictive and toward meaningful and expressive rituals.

The first quotation—"A man who reviews the old so as to find out the new is qualified to teach others"—is exactly the kind of review that the previous chapters have emphasized. For example, language is as old as the human race, and I have emphasized the importance of language throughout these pages. Analyses of the ideas of Plato and Nietzsche have taken us into the history of Western philosophy. By looking to Ecclesiastes, the ancient Indian philosophy or Hinduism, Buddhism, and Confucian ideas, we are also going far back into history. Similarly, our approach to social science does not neglect founders like Marx and Simmel. Confucius continues with this theme of the importance of the past in the second and third quotations, looking backward to what we have learned over time, to the experiences we have shared with friends, and also to the full range of what an individual has achieved throughout life. Does this orientation to the past prevent us from changing? Apparently not, for Confucius is maintaining that we "find out the new" when we review the old. In that way we go "back to the future."

Our analysis of rituals takes us into the past, that ocean of experiences we all have while swimming on its surface. Yet some rituals—which are our habits of action as distinct from habits of thinking or feeling—are relatively addictive, and others are relatively meaningful and expressive. The ideas of Confucius can help us to understand the nature of this difference. Confucius moved about from place to place during his lifetime, serving as advisor, tutor, and teacher to princes, governmental officials, business people, and others, focusing on people's experiences in everyday life. By contrast with Buddhism's emphasis on emotional detachment, Confucius emphasized jen or human-heartedness. For Confucius, loving your fellow human beings is the essence of virtue. And knowing others is the essence of knowledge. Relationships with others involving both knowledge and love illustrate meaningful and expressive rituals. This contrasts with the lack of such relationships to be found among individuals obsessed by some particular addiction or compulsion.

Such rituals require the individual to maintain a close relationship between aspirations and their fulfillment, by contrast with the large aspirations-fulfillment gap of contemporary society illustrated in Figure I-1.

> The moral life of man may be likened to traveling to a distant place: one must start from the nearest stage. It may also be likened to ascending a height: one must begin from the lowest step.... The path of duty lies in what is near, and men seek for it in what is remote ... is virtue indeed so far off? I crave for virtue, and lo! it is at hand. (quoted in Kaplan, 1961: 276)

Just as rituals have to do with behavior that is repeated, so is it a long-term commitment of the individual to jen or human-heartedness that a moral life requires:

> The way in which a man loses his proper goodness of mind is like the way in which the trees are denuded by axes. Hewn down day after day,

> can it—the mind—retain its beauty? But there is a development of its life day and night, and in the calm air of the morning, just between night and day, the mind feels in a degree those desires and aversions which are proper to humanity. But the feeling is not strong, and it is fettered and destroyed by what takes place during the day.... If it receives its proper nourishment, there is nothing which will not grow. If it loses its proper nourishment, there is nothing which will not decay. Hold it fast, and it remains with you. Let it go, and you lose it. (286-287)

Confucius's ideas here are similar to the idea of karma, as discussed in chapter 4. It is the individual's long-term actions that determine jen for Confucianism and karma for Hinduism and Buddhism. Our short-term actions can narrow the individual's behavior with a focus on one kind of addiction or another, reducing the karma he or she develops and denuding the individual's heart and mind in Confucius's terms. Yet the agricultural figures of speech used by Confucius are vague. Exactly what can we moderns do to develop meaningful and expressive rituals and hold fast to them rather than abandon them?

One direction that we can take is to dig deeper into an understanding of the history of Western philosophy in addition to Eastern religion, for that can help us to understand how we can learn to use the scientific method in our everyday lives and thus affect our worldview and aspirations-fulfillment gap. To illustrate, we might distinguish among four different approaches to what philosophers call the "dialectic": that of Socrates, of Plato, of Hegel, and of Marx. Socrates' orientation was to engage someone in a conversation or dialogue, usually centering on that individual's defining some concept (like justice). By bringing up examples that challenged that definition, Socrates succeeds in gaining adherence to his own definition. As for Plato, he was able to follow his teacher, Socrates, by achieving a synthesis of the ideas of two earlier philosophers, Heraclitus and Parmenides. Heraclitus centered on the idea that change rules the world: "One cannot step twice into the same river." Parmenides, by contrast, saw reality as permanent and unchanging. Plato achieved a synthesis of their ideas by seeing change in the material world and permanence in the world of ideas.

For Hegel, by contrast with Socrates and Plato, there is no conclusion that ends a dialogue once and for all—no final truth that we can cling to forever—for the dialogue continues. There is a "thesis" or an initial argument, an "antithesis" or a counterargument, and a "synthesis" that then becomes a thesis within a new argument. And the dialogue continues without ever reaching any final conclusion. And, for Hegel, this "triadic theory" covers far more ground than a dialogue between individuals, for it describes changes in ideas and in society generally throughout history. Marx was also interested in the grand sweep of history, and he also saw a triadic theory at work. But his emphasis was not on ideas, as in the case of Hegel, but rather on more tangible or material phenomena. He saw continuing contradictions between the "forces of production" (illustrated by the productivity of modern technology) and the "relations of production" (illustrated by the stratified or hierarchical relationship between owners and workers). The result, for Marx, was always a new synthesis that in turn became a thesis within the next stage of history. Yet the dialectic ends for Marx with the victory of the workers over the owners as a result of a revolution.

If we feel free to select ideas from these philosophers, by contrast with an all-or-none approach to each of them, then there is a great deal that we can learn from each of them. For example, from Socrates we can learn the importance of conversations with others about matters that are fundamental but without adopting the idea that it is our own view that will always prevail. Instead, we can work toward both individuals within a dialogue learning something important. Further, we can learn to engage in such dialogues as a genuine habit of action or a ritual, where we continue to learn from our interactions with others. From Plato, we can learn to integrate ideas from diverse sources without engaging in personal dialogues, just as I am attempting in these paragraphs from the work of Socrates, Plato, Hegel, and Marx, and just as I am attempting in this book as a whole. To accomplish this, one needs a very broad framework, as illustrated by that depicted in Figure I-3.

Thus, we can learn to relate Eastern and Western philosophy and religion to our rituals in everyday life. But these general ideas about altering our rituals require concrete illustrations if they are to become more meaningful to us. To this end, I will give some personal examples based on my own knowledge of and experiences with C. Wright Mills, the sociologist discussed in the introduction, who set the stage for my own approach to the scientific method. For example, it was his broad approach to social science, expressed in *The Sociological Imagination* (1959), that is the basis for Figure I-3. He followed the ideals and optimism of the eighteenth-century Enlightenment era in his commitment to the possibilities of human beings to solve society's fundamental problems. He saw the possibilities of a very broad approach to social science for providing the understanding that is desperately required for such solutions. And he understood the importance of communicating those ideas to people in general, versus hiding within the ivory tower of the academic world. He published books and magazine articles that were read by social scientists, but they were also read by a great many others.

Born on August 28, 1916, in Waco, Texas, Mills lived a relatively isolated life in his early years and continued to be a "loner" throughout his life (Phillips, 2004; see also Kathryn Mills, 2000). His father, an insurance broker, was away on trips much of the time, the family moved from place to place, and he developed no close and continuing relationships as he grew up. He entered Texas A&M, which was then a military school, and he soon became appalled by the hazing and general oppression of freshmen by upperclassmen. When he wrote a letter in protest to the student newspaper, he was accused in a rejoinder of a lack of "guts," and he responded in a second letter:

> Just who are the men with guts? They are the men who have the ability and the brains to see this institution's faults, who are brittle enough not to adapt themselves to its erroneous order—and plastic enough to change if they are already adapted; the men who have the imagination and the intelligence to formulate their own codes; the men who have the courage and the stamina to live their own lives in spite of social pressure and isolation. These my friends, are the men with "guts." (K. Mills, 2000: 34)

Mills transferred the next year, 1935, to the University of Texas with its strong departments of philosophy and sociology and its president a supporter of Roosevelt's New Deal. Mills did extremely well in philosophy, sociology, cultural anthropology, social psychology, and economics. He earned a master's degree in philosophy and then moved on to the doctoral program in sociology at the University of Wisconsin, having already published articles in the two leading sociology journals. Working as a teaching assistant in sociology, he developed a friendship with Hans Gerth, a European sociologist, and became interested in the classical European sociology of Max Weber and Karl Marx. Mills completed his course work in 1941 and then took a position in the Department of Sociology at the University of Maryland. In 1945 he became a research associate at Columbia University's Bureau of Applied Social Research, and a few years later became a professor of sociology at Columbia College, where he remained until his death in 1962.

Mills's doctoral dissertation, later published as *Sociology and Pragmatism: The Higher Learning in America* (1964)—centering on the work of Charles Peirce, William James, and John Dewey—carried forward his work on the philosophy of pragmatism as a student at the University of Texas. Mills became widely known for works that built on the classical sociological ideas of Weber and Marx and centered on patterns of social stratification, social class structures, and the alienation of the individual. His trilogy—*The New Men of Power: America's Labor Leaders* (1948), *White Collar: The American Middle Classes* (1951), and *The Power Elite* (1956)—was addressed not just to academic sociologists but also to intellectuals in general, and it was widely read and translated into many languages. Together with Hans Gerth, who had helped to steer Mills's interests in this direction, he edited a well-known collection of essays, *From Max Weber* (Gerth and Mills, eds., 1946), and wrote *Character and Social Structure: The Psychology of Social Institutions* (Gerth and Mills, 1953). Later publications—including many articles in liberal magazines—were based on efforts to influence major political decisions in the United States: *The Causes of World War Three* (1958); *Listen, Yankee* (1960), based partly on interviews with Castro, Che Guevara, and others during his visit to Cuba; and his last book, *The Marxists* (1962).

Yet even more than Mills's publications on social stratification, his work on the scientific method appears to have had the most profound and long-lasting impact on contemporary social science. We see that interest of Mills appear early in his education, with his study of Peirce, Dewey, James, and George Herbert Mead as an undergraduate and as a candidate for a master's degree in philosophy at the University of Texas. We may note it also in his doctoral dissertation on sociology and pragmatism, a philosophy that provides a framework for the scientific method. And we can see it in one of his earliest articles, "The Professional Ideology of Social Pathologists" (1943), where he critically analyzes textbooks on social problems:

> The level of abstraction which characterizes these texts is so low that often they seem to be empirically confused for lack of abstraction to knit them together. They display bodies of meagerly connected facts, ranging from rape in rural districts to public housing, and intellectually sanction this low level of abstraction.... Collecting and dealing in a fragmentary way with scattered problems and facts of milieux, these books are not focused on larger stratifications or upon structured wholes.... (1943: 168)

It is this idea of moving from general or abstract concepts like "social stratification" to specific or more concrete concepts—like "sexism," "classism," "ageism," and "ethnocentrism"—that is much of the basis for the ability to integrate social science knowledge, as presented in the introduction to this book. Unfortunately, however, this quote illustrates not simply the nature of the textbooks on social problems that Mills analyzed but also contemporary research throughout the social sciences. In *The Sociological Imagination* (1959) Mills spelled out more thoroughly than ever before his vision of a scientific method that can work for social science. That method included use of a "sociological imagination" broad enough to encompass the complexity of human behavior, as quoted in the introduction:

> The sociological imagination ... is the capacity to shift from one perspective to another—from the political to the psychological; from examination of a single family to comparative assessment of the national budgets of the world; from the theological school to the military establishment; from considerations of an oil industry to studies of contemporary poetry. It is the capacity to range from the most impersonal and remote transformations to the most intimate features of the human self—and to see the relations between the two. (1959: 7)

This was the book that inspired the members of the International Sociological Association to rank it as the second most influential book for sociologists throughout the entire twentieth century. In that book Mills developed further his idea as to the importance of the linguistic distinction between abstract and concrete concepts, which he had written about in his 1943 article "The Professional Ideology of Social Pathologists":

> One great lesson that we can learn from its systematic absence in the work of the grand theorists is that every self-conscious thinker must at all times be aware of—and hence be able to control—the levels of abstraction on which he is working. The capacity to shuttle between levels of abstraction, with ease and with clarity, is a signal mark of the imaginative and systematic thinker. (1959: 34)

It is this ability "to shuttle between levels of abstraction" that is a fundamental part of the scientific method, for it enables the social scientist to integrate diverse concepts and diverse bodies of research that would otherwise remain hidden among the many specialized and subspecialized areas throughout the social sciences. And that integration enables the social scientist to confront the complexity of human behavior.

I had an opportunity to observe Mills's sense of problem or critical orientation firsthand as a student at Columbia College in the 1950s, where his courses influenced me to change my major from premed to sociology. I recall watching from an upstairs window as Mills came roaring into the College's quadrangle on his motorcycle, with the tail of his coonskin cap waving in the breeze. He was both tall and wide, and he made the most of his physical presence in the classroom, with a voice that was deep and highly expressive emotionally. It is this kind of commitment or habit of expression that makes rituals or habits of action emotionally expressive and that is also fundamental to practicing the scientific method.

If we now look to Mills's habits of action or ritualistic behavior, once again we can see continuities throughout his life. He was someone who relished challenging existing authorities. We can see this approach in his letter to the student newspaper at Texas A&M where he wrote that men with guts are "the men who have the imagination and the intelligence to formulate their own codes; the men who have the courage and the stamina to live their own lives in spite of social pressure and isolation." Mills was proud of what he did in an encounter with Dwight Eisenhower, who was then president of Columbia University. Mills told his class that one day Eisenhower walked in, unannounced, and sat down in the back row. Mills immediately changed his topic to discussing how the class could act as a cell directing a violent revolution to overthrow the U.S. government. Eisenhower's reaction was to become more and more uncomfortable, and he finally left the room and was never heard from afterward. This was an oppositional stance that—throughout his life—generally kept Mills from joining with others to achieve common objectives. On the one hand, it served him well in his feeling free to advance his own ideas. On the other hand, however, this "loner" approach left him with no group working with him and helping him to further develop and communicate his ideas.

Although Mills's work throughout his life coupled with my contact with him provided the basis for my own efforts to develop a scientific method that can address the full complexity of human behavior, it is crucial that we learn to carry further his legacy rather than worship his ideas. That is the way the scientific method should work. This book and earlier books as well as the work of the Sociological Imagination Group—as discussed in the introduction—are efforts to build on Mills's work along with that of a great many others. As one illustration of these efforts, I and a colleague in the Sociological Imagination Group—David Christner—have developed what we call "The Evolutionary Manifesto," a 5-page document excerpted here and available on the Web site www.sociological-imagination.org. We may ask to what extent this manifesto succeeds in carrying further the meaningful and expressive rituals centered on the scientific method that Mills manifested during his life.

The Evolutionary Manifesto

We—Bernard Phillips and David Christner—two of the founding members of the Sociological Imagination Group, are committed to the integration of existing knowledge of human behavior and its application to world-wide problems.

Problems

1. We are convinced that problems throughout the world are accelerating, and that they are beginning to threaten the very survival of the human race. We see the 20th century as by far the bloodiest century in history, with its wars and deliberate murder of millions of civilians yielding the slaughter of more people than in all previous wars. And we are aware of no systematic direction within any nation for either understanding the nature of these problems or confronting them effectively, granting partial successes. Yet we are also aware of the infinite potential of every single human being. This is a time neither for unrealistic pessimism nor for unrealistic optimism, but rather for a realistic understanding of our complex and threatening problems and our potential for confronting them effectively....

Recommendations

Deep Democracy. We see a broad approach to the scientific method as yielding the basis for a long-range goal of a much deeper form of democracy than has yet been achieved anywhere on the globe. It was Jane Addams who claimed in her *Democracy and Social Ethics* (1902) that "The cure for the ills of democracy is more democracy." Those ills, as illustrated by widespread patterns of social stratification which point far away from egalitarian ideals, are evident in the United States and Western Europe as well as in other nations. Our first recommendation, then, is that a long-range goal for all peoples should be the continuing extension of democracy far beyond political processes and into the everyday behavior of individuals and groups. During the 1960s the "countercultural" movements in the United States pointed toward the achievement of more equality throughout society without emphasizing the importance of gaining further understanding of human behavior by building on the achievements of the academic world. We believe that they failed to achieve their aims largely as a result....

The Scientific Method in Everyday Life. If the scientific method is to help all of us move toward deep democracy, then we should all have as a fundamental goal our movement toward learning to use a broad approach to the scientific method in our own everyday lives. We recommend, then, that using a very broad approach to the scientific method in our everyday lives should be a priority for us that is no less fundamental than our making the development of deep democracy a priority.

We are all in different situations, and as a result different individuals will work in different ways if they follow this recommendation. For example, a political leader might wish to emphasize legislation that would fund social scientists to work toward the integration of their knowledge. Some might choose to read social science publications. Others might begin to focus on examining their own lives. People working with mass media might attempt to develop films or novels which illustrate or carry further the above ideas. Those deeply involved with computers might choose to explore how individuals can learn to integrate the vast bodies of knowledge presently available on the internet. Social scientists and others might explore the ideas of the Sociological Imagination Group—perhaps beginning with its website www.sociological-imagination.org or participating in its annual conferences or corresponding with us—in order to develop further or challenge those ideas. Journalists working in print or on television may choose to write about these ideas of deep democracy and the scientific method in everyday life. Physicians may explore the possible impact of these ideas on the health of the individual. Attorneys might examine the potential of these ideas for improving the effectiveness of legal systems as well as their personal effectiveness in their practices.

The previous section of the manifesto on problems carries forward Mills's concerns with problems such as war and social stratification into our present era. Given our more devastating weapons at this time in history, with no solutions being put forward by political leaders, those problems have escalated since Mills's time. As for the recommendations, the manifesto carries forward Mills's democratic orientation—given his severe criticism of social stratification—and also follows his interests in communicating to people in general and not just to academicians.

With this chapter we complete our introduction to personality structures, preparing the way to examine social structures in part 4. Mills was much concerned with the negative impact of social structures on the individual, and he succeeded in helping us to understand how that impact actually works. What we require at this point is an understanding of social structure that takes the possibilities of the individual very seriously, versus the specialization that generally keeps these two entities apart from one another. For example, there is the split between those social scientists emphasizing the top curve of culture and personality within Figure I-1 and those focusing on the bottom curve of social organization. Mills attempted to make progress on this in his book coauthored with Hans Gerth, *Character and Social Structure* (1953). Given what we've learned since Mills's time, we are now in a position to carry further those efforts. What is required is a very broad approach that also emphasizes the ways in which the abstract concepts of social science are linked together. For example, we will be concerned not just with rituals but also with social rituals, and not just with values but also with cultural values. In achieving this, we will continue to be guided by our "heart," "head," and "hand" figures of speech, for they can help us to see more clearly how those abstract concepts are related to one another.

PART IV

Social Structures

Before proceeding to chapters 6, 7, and 8, I would like to present the key concepts invoking social structures to be examined in those chapters. Continuing our analysis of social science's extraordinary language, Figure 6-1 combines Figure 4-1's focus on personality structures with concepts centering on social structures. Examining Figure 6-1, we might note that all of these concepts are defined explicitly in the glossary. And all of them deal with structures, by contrast with the situational emphasis in vernacular or everyday speech. We shall add concepts—which are themselves structures—that bear on situational phenomena at the beginning of part 5, the "Situation," as indicated in the center of Figure 9-1. Further, almost all of the structures referred to in Figure 6-1—the exceptions are the three concepts in boldface, namely, individual, society, groups—are relatively invisible, by contrast with the structures we see all around us. Although the concepts in Figure 6-1 help us to become aware of their existence, it is most difficult for us to retain that awareness as we move through life. The major structures are capitalized: physical, personality, biological, and social structures. The italicized concepts within personality structure (values, beliefs, assumptions, rituals) and within social structure (cultural values, cultural norms, social rituals) are key elements of those structures, with other aspects of personality and social structures building on those elements.

Although all of the concepts in Figure 6-1 are linked to one another, those links between personality and social structures that are particularly important are numbered. The same number is used for any given pair of concepts, with an apostrophe added to the concept from social structures in any given pair. Beginning with the italicized concepts, "social rituals" (3') are simply personal rituals that are shared within some group, such as a family's pattern of celebrating its children's birthdays with a party. But

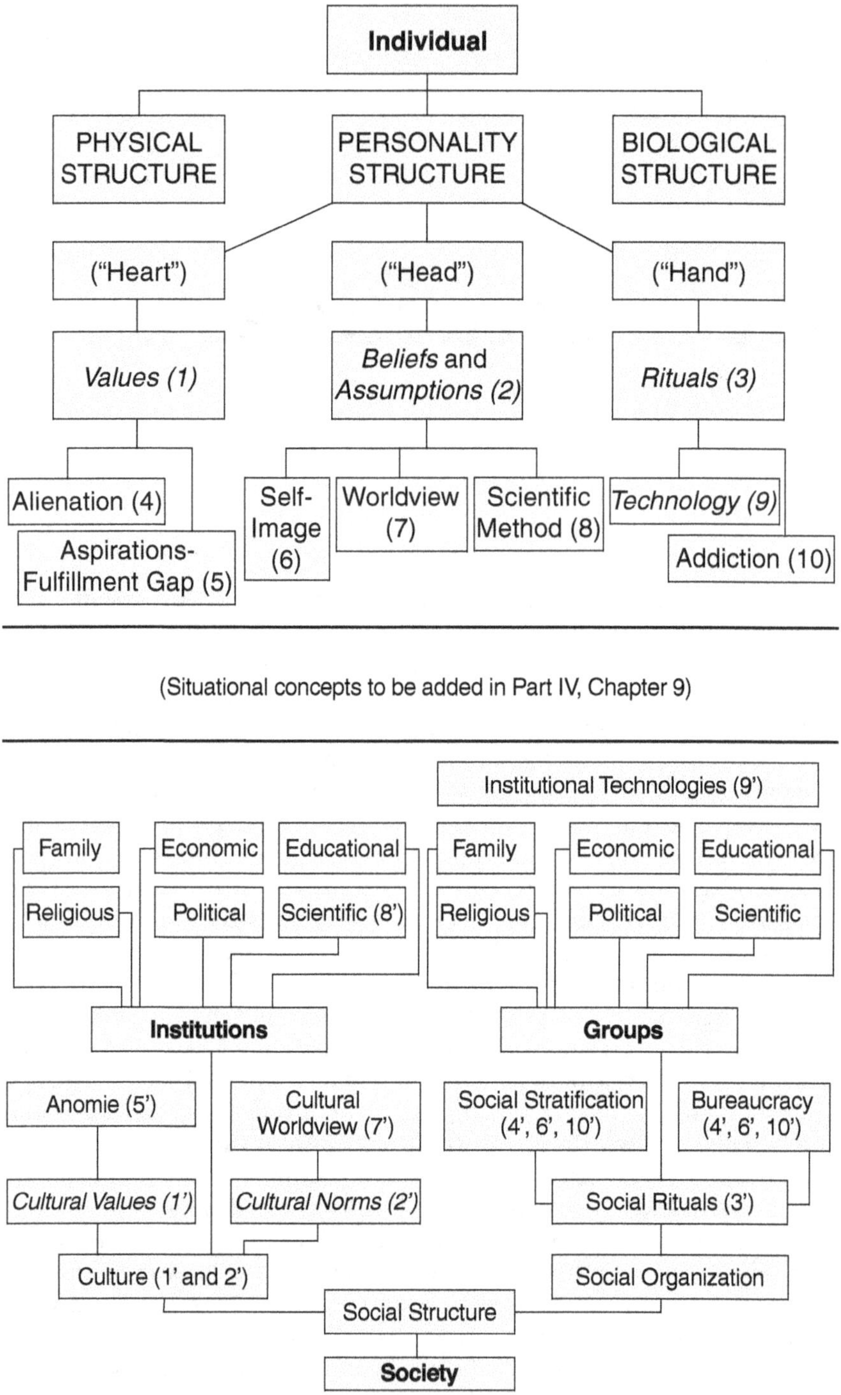

Figure 6-1. Relationships among Basic Structures within the Individual and Society

in the case of "cultural values" (1') and "cultural norms" (2')—corresponding to the values (1) and beliefs and assumptions (2) of an individual—the group involved is an entire people, such as the annual Thanksgiving Day celebration in the United States. Cultural values and norms, thus, are very widely shared among a people, corresponding to the breadth of "culture" itself, granting the many instances of deviant behavior. This pairing of concepts from social and personality structures is most significant for our understanding of human behavior. If personality structures are intangible, then social structures are much more invisible, yet they remain the most powerful forces in shaping our personality structures. For example, to the extent that we learn the nature of our cultural values we come to understand a great deal about the nature of the values held by any single individual participating within that culture, granting that individuals also have values that are unique to their own experiences. And the same holds true for cultural norms. Social scientists generally have not emphasized the educational potential of such knowledge of culture because of their own specialized orientations. In chapters 6, 7, and 8 we will examine the linkages depicted within Figure 6-1 in some detail, by contrast with our brief sketch in this introduction to part 4.

The link between "anomie" (5') and "aspirations-fulfillment gap" has never been made—as far as I know—within the social science literature, yet it seems to yield substantial insight into the nature of anomie as well as the broad implications of the aspirations-fulfillment gap. The same is the case for "cultural worldview" (7') and "worldview" (7). Granting some uniqueness in the fundamental assumptions or metaphysical stance that each of us adopts, there is little question but that our worldview is largely shaped by the worldview predominating within our culture. And once again, learning the nature of that cultural worldview takes us very far toward understanding the nature of the worldview of any given individual within that culture. The pairing of our personal approach to using the scientific method (8)—following Kelly's argument from chapter 4 that we might all do well to see ourselves as scientists—with the definition of the scientific method within the scientific institution (8') gives us a direction for building on Kelly's insight. Following the general argument in this book, we may choose to develop our efforts to use the scientific method in our everyday lives to reach a greater understanding of our world and ourselves, and to address fundamental world and personal problems. As for the actual procedures or technology (9) we use to solve problems, the link between those procedures and the technologies (9') in place within our various institutions can also help us in much the same way, suggesting directions for improving our own personal technologies. The remaining links between personality and social structures depicted in Figure 6-1 are not so straightforward. Three personality structures (alienation: 4, self-image: 6, and addiction: 10) are paired with each of two social structures (social stratification: 4', 6', and 10'; and bureaucracy: 4', 6', and 10'). This effort—to be explained in the following chapters—points toward seeing personal problems like alienation and addiction as tied closely to the social structures of social stratification and bureaucracy.

As we proceed to part 4 let us bear in mind the importance of our learning to link personality structures with social structures. This is a fundamental lesson of the social sciences: that our own personal situation can be understood far more clearly by looking to the social forces that shape that situation. Further, an awareness of the nature of those forces can help us to understand and move toward solving personal and world problems. This lesson from the social sciences requires us to make a leap into the void, for those social forces are largely invisible. It also requires us to challenge personal beliefs and assumptions that conflict with social science knowledge. We require a willingness to assess that knowledge on the basis of our personal experiences, just as scientists proceed to test new hypotheses. We might recall Peirce's idea from chapter 4: "The irritation of doubt is the only immediate motive for the struggle to attain belief." Given increasing world problems coupled with "the invisible crisis of contemporary society"—based on Figure I-1's growing aspirations-fulfillment gap—the time has come for us to question our basic assumptions about society and ourselves that previously we have avoided questioning. Following Peirce, we must learn to doubt what we had previously taken for granted. We can then proceed by following Peirce's dictum: "Do not block the way of inquiry." We are all scientists to a degree, and we can continue to learn—without any limit whatsoever—how to understand ourselves and the world, and how to solve our problems. This requires, however, that we become scientific to a greater extent than previously. Chapters 6, 7, and 8 will help us learn to build on social science's understanding of social structures, and how they are linked to personality structures. Chapters 9, 10, and 11 will help us learn how to understand the complexities of the momentary situation. All six chapters will carry further both our understanding of how to develop the ability to make use of more of the potential that language offers us and how to apply the scientific method to our own everyday experience.

CHAPTER 6

"Heart"

Anomie Versus Cultural Value Fulfillment

IN OUR EFFORT TO EXAMINE LINKS AMONG THE CONCEPTS DEPICTED in Figure 6-1, let us turn to C. Wright Mills's view of the fruits of a sociological imagination:

> Perhaps the most fruitful distinction with which the sociological imagination works is between "the personal troubles of milieu" and "the public issues of social structure." ... In so far as an economy is so arranged that slumps occur, the problem of unemployment becomes incapable of personal solution. In so far as war is inherent in the nation-state system and in the uneven industrialization of the world, the ordinary individual in his restricted milieu will be powerless—with or without psychiatric aid—to solve the troubles this system or lack of system imposes upon him. In so far as the family as an institution turns women into darling little slaves and men into their chief providers and unweaned dependents, the problem of a satisfactory marriage remains incapable of purely private solution. In so far as the overdeveloped megalopolis and the overdeveloped automobile are built-in features of the overdeveloped society, the issues of urban living will not be solved by personal ingenuity and private wealth. (1959: 8, 10)

We can look to "personal troubles" as an aspect of personality structure, and we might see "public issues" as an aspect of social structure. Just as Figure 6-1 emphasizes the links between personality and social structure, so does Mills emphasize those links by pairing personal troubles with public issues. A major point of Mills as well as the discipline of sociology is that generally we fail to be aware of these links. For example, the individual who loses her or his job generally blames self, not taking into account the power of social structures—which remain largely invisible—in creating that job loss. And so it is with the range of other "personal troubles" that we individuals experience.

Yet it is not just social structures that we are often unaware of, as sociologists generally emphasize. We are often unaware of our personality structures, as psychologists generally emphasize. We have already noted such an outward emphasis in our worldview in chapter 2, where we looked outward rather than both inward and outward. As a result, we failed to take personal responsibility for problems, blaming city hall rather than looking to our own behavior. This approach was well illustrated by Vaclav Havel in his inaugural address as Czech president in 1990:

> We live in a contaminated moral environment. We fell morally ill because we became used to saying something different from what we thought ... in other words, we are all—though naturally to differing extents—responsible for the operation of the totalitarian machinery; none of us is just its victim: we are all also its co-creators ... we have to accept this legacy as a sin we committed against ourselves. If we accept it as such, we will understand that it is up to us all, and up to us only, to do something about it. We cannot blame the previous rulers for everything, not only because it would be untrue but also because it could blunt the duty that each of us faces today, namely, the obligation to act independently, freely, reasonably and quickly.... Freedom and democracy include participation and therefore responsibility from us all. (quoted in Montefiore, 2005: 212–213)

Apparently it is the full range of structures depicted in Figure 6-1, and not just social structures, that we often fail to see in our everyday efforts to understand human behavior and solve our problems. Instead of becoming aware of structures, which would take us beyond our momentary situation, we remain in that momentary situation to a great degree, much like forms of life without language. Following chapter 1, we remained in Flatland rather than Timeland and Spaceland, even though our language gives us the potential to move out of Flatland. With respect to chapter 2, we focused on the physical phenomena surrounding us within any given momentary scene, rather than perceiving the invisible social and personality structures that could take us far beyond that scene. As for chapters 3, 4, and 5, we failed in particular to become aware of our growing aspirations-fulfillment gap, and as a result, we remained unable to take actions that would narrow that gap. Overall, then, we remained victims of forces that we failed to see and that increasingly threaten us. We have the linguistic capacity to detect those forces and to reduce those threats, but we have yet to learn how to translate that capacity into the ability to initiate decisive actions to meet those threats effectively.

Our initial focus in this chapter is on those social structures, emphasizing the heart or emotions, as well as on their links with the personality structures examined in previous chapters, as illustrated by the phenomenon of anomie. The concepts to be introduced are part of what Gouldner called the extraordinary language of social science. Those concepts help us to become aware of the invisible social forces that continue to shape our lives. The final section of the chapter centers on further movement toward confronting those social forces to narrow our aspirations-fulfillment gap, as illustrated by the idea of cultural value fulfillment. In this section our concern is with learning to move out of our Flatland situation so that we can see the social and personality structures—such as our aspirations-fulfillment gap—that shape our lives from one moment to the next. In that way we can move toward learning how to narrow a gap that increasingly threatens all of us at this time in history.

Anomie

Let us recall the quote from Durkheim's *Suicide* in the introduction:

> Such is the source of the excitement predominating in this part of society [trade and industry] and which has thence extended to the other parts. There, the state of crisis and anomy is constant and, so to speak, normal. From top to bottom of the ladder, greed is aroused without knowing where to find ultimate foothold. Nothing can calm it, since its goal is far beyond all it can attain. (Durkheim, 1897/1951: 254–255)

Durkheim refers here to the industrial revolution's encouraging of a materialistic aspiration "far beyond all it can attain," linked to the phenomenon of "anomy" or "anomie," which literally means normlessness. With regard to the upper portion of Figure 6-1, Durkheim is invoking the materialistic aspirations or "values" of the individual, a key aspect of the individual's personality structure. Yet he is also invoking the lower portion of the figure that has to do with social structures, for those individual values did not suddenly arise out of thin air. They emerged as a result of an industrial revolution that yielded values widely shared throughout society—or "cultural values"—emphasizing the goal of the increasing acquisition of goods and services. "Anomie" is thus an element of social structure that in turn influences the development of an element of personality structure, just as cultural values influence the development of individual values.

But there is more to the concept of anomie than this. Durkheim states that the goal of this "greed" that "is aroused" by the industrial revolution "is far beyond all it can attain." From the perspective of the individual, he is referring to a gap between aspirations or values and their fulfillment, which is listed in Figure 6-1 as a personality structure. Yet when we begin to see that substantial aspirations-fulfillment gap as one that is widely shared throughout society—moving to the lower half of Figure 6-1—we may refer to that gap as "anomie." Here, then, I am defining anomie as substantial gaps between aspirations and their fulfillment that are widely shared throughout society. Dare I do this, given that this definition differs not only from that of Durkheim but also from any of the definitions of anomie that have been put forward throughout the sociological literature? Absolutely! Definitions are tools that may or may not be able to help us understand phenomena and solve problems. As such, they are all most provisional, to be assessed on the basis of how fruitful they prove. To depart from definitions by key figures in the discipline may appear to some to be blasphemy. Yet the scientific method has nothing whatsoever to do with the worship of the founding fathers. By contrast, if a definition succeeds in clarifying relationships between personality and social structures—a most difficult task throughout the social sciences—then more power to it.

There is yet more to the concept of anomie, for this concept not only points up aspirations-fulfillment gaps throughout society but also suggests that individuals throughout society remain unable to close that gap. In other words, anomie is a social structure in that it persists over time. The gap depicted in Figure I-1 remains there and presumably continues to increase. This is a very serious matter for society as a whole, for it implies that, generally, people throughout society are failing to fulfill their own fundamental goals. Durkheim's emphasis on suicide suggests just how serious these gaps can be. To begin to understand the sources of anomie we must bring to bear not only our understanding of culture—or beliefs and values widely shared throughout society—but also our understanding of "social organization," located on the right side of the lower portion of Figure 6-1. Although we defer this topic until chapter 8, where we shall take up concepts having to do with "hand," a few words might be said here. It is patterns of action like "social stratification" and "bureaucracy" that appear to stand in the way of cultural value fulfillment. For those phenomena are relatively rigid, contrasting sharply with the scientific method. As a result, they generally prevent the individual from learning how to solve a given problem, such as an aspirations-fulfillment gap.

Although the concept of anomie can refer to materialistic values, it is broad enough to invoke nonmaterialistic values as well. For an example we can examine Vaclav Havel's inaugural address, where he referred to freedom and democracy as well as "the totalitarian machinery." We have only to bring forward our sociological concepts to understand more clearly what he is saying. Freedom and democracy are the values of not only particular individuals in Czech society and elsewhere, they are also very widely shared values of individuals throughout society, or "cultural values." Havel and others throughout Czech society were much concerned with the gap between these individual and cultural values and the ability of individuals to fulfill them. In other words, they saw Czech society as in a state of "anomie," characterized by widely shared aspirations-fulfillment gaps. And we might see "the totalitarian machinery" that Havel cited as referring to widespread patterns of "social stratification" and "bureaucracy," which were largely responsible for preventing people from narrowing their aspirations-fulfillment gaps or reducing the anomie throughout Czech society.

For further understanding of the persistence of anomie in contemporary society, we can look to the "head" as well as the "heart" and the "hand," previewing what will be examined more thoroughly in chapter 7. In our examination in chapter 1 of the forces keeping the individual in Flatland and preventing her or him from moving into Spaceland and Timeland, we noted two intellectual orientations. "The technique of particularization" is an individual's pattern of beliefs used to maintain a narrow focus while avoiding the threats that might result from a broader orientation, as illustrated by the Springdalers's centering on the attributes of a particular wealthy individual and avoiding acknowledgment of the difference in social class between her- or himself and those who are wealthy. We might also think of highly specialized social scientists who focus on the contributions of their own work rather than on their failure to follow the ideals of the scientific method by opening up to the relevant discoveries of other social scientists in neighboring fields of knowledge. A second intellectual orientation or pattern of belief is "the falsification of memory." For example, a farmer, "instead of entertaining the youthful dream of a 500-acre farm, entertains the plan to buy a home freezer by the fall." Or a sociologist, instead of thinking of the belief he adopted in midcentury of solving fundamental problems in society, trades that earlier belief for a belief in his ability to become a full professor in his sociology department.

A way of looking at such beliefs, advanced by the *Boston Globe* columnist James Carroll, is that we moderns all live within a "culture of self-deception," and it is that very culture that prevents us from becoming aware of such problems as an individual's aspirations-fulfillment gap or a culture's anomie:

> "My dear fellow citizens," Vaclav Havel said in his inaugural address as Czech president, "for 40 years on this day you heard from my predecessors the same thing in a number of variations: how our country is flourishing, how many millions of tons of steel we produce, how happy we all are, how we trust our government, and what prospects lie ahead for us. I assume you did not propose me for this office so that I, too, should lie to you."...
>
> The pattern is wide. Executives who want only to put the numbers "in a better light" end by cooking the books. Politicians who harmlessly aim to tell voters what they want to hear wind up having no core grasp of what is true. Religious leaders who maintain the appearance of virtue as an absolute value lose the capacity to recognize their own fallibility.... Thus, lifetime partners can go years without realizing they have no intimacy. The overweight can fool themselves about their health problems. Drinkers can deny what their lives have become. Compulsive workers can

> enslave themselves in a false dream of success. Life-wrecking depression can pass itself off as selfless worry. Greed can seem like ambition. The pursuit of happiness is killing us. The most damaging lies are the ones we tell ourselves. . . .
>
> Today's nationwide rude awakening understandably prompts broad outrage at deceptive leaders, but that must not be our only response. During the Cold War we exempted ourselves from the kind of moral scrutiny that could have prevented the spreading of this poison cloud. The lies of business (Madison Avenue), of government (the Missile Gap), of culture (Forever Young), and of religion (God Bless America) were built into our system, but we could not see them as lies because of the split in our thinking. In this time of ethical reckoning, will we again smugly divide the world between the good and the evil?
>
> Yes, we can insist on a purging of public lies—a true reform. But we repeat a major blunder if we conclude that the problem of self-serving deception belongs to someone else. "When I talk about the contaminated moral atmosphere," Havel presumed to tell his fellow citizens, "I mean all of us." (Carroll, 2002: A15)

Apparently, this "culture of self-deception" that Carroll writes about is extremely widespread. We might think here of Socrates with his advice that "the unexamined life is not worth living." Or we might think of Walt Kelly's cartoon-strip character, the opossum Pogo, who looked at the litter deposited under a tree and said: "We have met the enemy and he is us," taking off on Commodore Oliver Hazard Perry's report of his victory over the British in the Battle of Lake Erie on September 10, 1813: "We have met the enemy and they are ours." This meshes with the idea put forward in chapter 2 that our worldview or metaphysical stance is outward-oriented, versus oriented both inward and outward. By contrast, we have Alvin Gouldner's idea of a "reflexive sociology," presented in the introduction (1970), and we also have "reflexive analysis" as an element of the Web and Part/Whole Approach to the scientific method, presented initially in chapter 1 and continued in subsequent chapters. Cultures are structures that are most difficult to change. And as structures, they persist from one moment to the next. Yet cultures can be changed, and so can worldviews or metaphysical assumptions. It is indeed possible to reduce our aspirations-fulfillment gap along with our degree of anomie. Carroll implies as much when he asks the question, "In this time of ethical reckoning, will we again smugly divide the world between the good and the evil?" And Havel pointedly calls on all individuals to take responsibility for changing their "contaminated moral environment" and working toward a better future for themselves and society: "Let us not be mistaken: the best government in the world, the best Parliament and the best President, cannot achieve much on their own. And it would also be wrong to expect a general remedy from them only. Freedom and democracy include participation and therefore responsibility from us all."

Another illustration of widespread anomie in contemporary society comes from a 2007 op-ed piece by Betsy Hart, author of *How the Culture of Pushover Parenting Is Hurting Our Kids—and What to Do about It.* "We Live in a Can't-Say-No Culture" centers on a new weight-loss pill—"Alli"—that the Food and Drug Administration has just approved for over-the-counter sales and that will be available in a matter of months. The problem of obesity is no minor one, for Hart indicates that "as many as 300,000 people a year may die prematurely because of obesity," making this "the second largest cause of preventable death, just behind smoking and closing in fast on becoming No. 1." Hart believes that the new drug not only will have zero impact on America's "bottom line" but might in fact make us even fatter as a nation:

> We are a nation of people who hate to say "no" to ourselves and who know less and less about delayed gratification. Whether it's food, sex, buying stuff, giving way to our "feelings" of the moment, it has become "all about me and what I want right now." And we Americans want a lot. Now.... [This is] a symptom of a can't-say-no culture. As such, the problems that have us Americans more overweight than ever are problems that no pill can really fix. (February 14, 2007: 13A)

What we have here is an illustration of the individual's large aspirations-fulfillment gap as well as of anomie as part of the culture of modern society. As Figure I-1 illustrates we are experiencing a continuing revolution of rising expectations—"whether it's food, sex, buying stuff, giving way to our 'feelings' of the moment"—and cultural values are going through the roof. This is the kind of thing that Durkheim was suggesting when he wrote that "greed is aroused without knowing where to find ultimate foothold." In attempting to fulfill our unrealistic expectations, however, our pattern of actions can yield such life-threatening problems as obesity. By learning to say no to unrealistic expectations—that is, by refusing to be caught up in escalating materialistic cultural values—we can lower our aspirations, narrow our aspirations-fulfillment gap, and work toward reducing the anomie prevalent throughout society. Pointing away from an outward-oriented worldview where we might look to a pill to solve an obesity problem, this approach points toward an inward-outward worldview where we take responsibility for our own behavior, following the orientations of Vaclav Havel and James Carroll. It is an approach that also follows the pragmatic strategy of lowering aspirations or "thirst" put forward by Buddha several thousand years ago (chapter 2) and advanced as well by Confucius (chapter 5). This orientation can become an East-West strategy to the extent that it is coupled with following our high Western aspirations in the long run, so long as we manage to keep anomie in check by making sure that our curves of aspiration and fulfillment—as depicted in Figure I-1—remain close to one another.

Although Durkheim focused on anomie in relation to society as a whole as well as the culture of society, we can extend that concept to include segments of society and subcultures. By so doing, we can use that concept to open up to more of the complexity found in society, especially since we are also viewing anomie as the shared aspirations-fulfillment gaps of individuals. This approach to anomie was recently taken up by a sociologist using the Web and Part/Whole Approach. In his analysis of terrorism, Adam Rafalovich viewed the concept of anomie as broad enough to include Durkheim's cultural orientation, Robert Merton's subcultural approach to deviance within his theory of structural strain, and Harold Garfinkel's situational orientation to deviant behavior (Rafalovich, 2007).

The key to understanding anomie—as well as the individual's aspirations-fulfillment gap—is the nature of cultural values, or those values widely shared throughout society. For it is those cultural values that succeed in shaping the values or aspirations of the individual along with the widely shared aspirations that form anomie to the extent that they remain unfulfilled. For example, by knowing the importance of the cultural value of equality, we also have a good start toward understanding the aspirations-fulfillment gap experienced by individuals throughout that society. Even if we've never met such individuals, this knowledge of cultural values yields insight into their own individual values. Yet how are we to learn the nature of cultural values, given the wide range of cultures to be found in the modern world? Anthropologists have taught us the diversity of cultures, and that each culture is unique and must be understood on its own terms.

Yet this emphasis of anthropologists appears to be one-sided, yielding a distorted picture of human beings, as argued by Donald Brown, an anthropologist who has made a thorough study of this situation:

> Although they were sent into the field with the charge of getting the whole picture, so that they could come back relieved of parochial views and thus tell the world what people are really like, anthropologists have failed to give a true report of their findings. They have dwelt on the differences between peoples while saying too little about the similarities (similarities that they rely upon at every turn in order to do their work).... The more those [sociocultural] differences can be shown to exist ... the more sociocultural anthropologists (or sociologists) can justify their role in the world of intellect and practical human affairs and thus get their salaries paid, their lectures attended, their research funded, and their essays read. (1991: 154)

Just as I have examined the failures of contemporary sociologists to follow the ideals of the scientific method, so does Brown examine the failures of contemporary anthropologists to "tell the world what people are really like." To focus on differences—whether between some long-dead tribe and us or between Nazi Germany and American society—relieves us of the responsibility of taking a hard look at our own problems. Instead, we can learn to feel comfortable about ourselves and avoid what appear to be crises that threaten to destroy our world, using "the techniques of particularization" and "the falsification of memory," as described by Vidich and Bensman (1960). Yet let us not be too harsh in pointing the finger of blame at contemporary social scientists while feeling good about ourselves. For we all appear to be victims of a stratified worldview or metaphysical stance that sharply limits our ability to follow our own ideals. Referring to Figure I-2 in the introduction, we can see that worldview as fundamental to shaping our aspirations-fulfillment gap—corresponding to anomie—as well as our basic problems. Our discussion of the importance of this worldview pervades all of the preceding chapters. Given our stratified worldview's ability to push us into techniques of particularization and the falsification of memory, however, it would be useful for us to examine another writer who communicates the importance of looking to metaphysics:

> Education cannot help us as long as it accords no place to metaphysics. Whether the subjects taught are subjects of science or of the humanities, if the teaching does not lead to a clarification of metaphysics, that is to say, of our fundamental convictions, it cannot educate a man and, consequently, cannot be of real value to society.... What is at fault is not specialization, but the lack of depth with which the subjects are usually presented, and the absence of metaphysical awareness.... The sciences are being taught without any awareness of the presuppositions of science, of the meaning and significance of scientific laws, and of the place occupied by the natural sciences within the whole cosmos of human thought.... Economics is being taught without any awareness of the view of human nature that underlies present-day economic theory.... How could there be a rational teaching of politics without pressing all questions back to their metaphysical roots?...
>
> We shrink back from the truth if we believe that the destructive forces of the modern world can be "brought under control" simply by mobilising our resources—of wealth, education, and research—to fight pollution, to preserve wildlife, to discover new sources of energy, and to arrive at more effective agreements on peaceful coexistence.... The "logic of production" is neither the logic of life nor that of society. It is a small and subservient part of both. The destructive force unleashed by it cannot be brought under control, unless the "logic of production" itself is brought under control—so that destructive forces cease to be unleashed. It is of little use trying to suppress terrorism if the production of deadly devices continues to be deemed a legitimate employment of man's creative powers.... (Schumacher, 1973: 86–87, 278–279)

Schumacher's ideas are not to be taken lightly, given his background as economic advisor to the British Control Commission in postwar Germany as well as his tenure as the top economist and head of planning at the British Coal Board for twenty years. His analysis of the failures of education to confront our fundamental metaphysical assumptions suggests the general failure of education to prepare contemporary society for escalating visible and invisible problems. His overall critique, however, centers on an economic system oriented to a materialistic way of life, which he contrasts with Buddhist economics: "The Buddhist point of view takes the function of work to be at least threefold: to give a man a chance to utilize and develop his faculties; to enable him to overcome his ego-centeredness by joining with other people in a common task; and to bring forth the goods and services needed for a becoming existence" (1973: 51).

Schumacher's focus on the failure of society's efforts to confront its fundamental problems takes us back to Martin Rees's *Our Final Hour* (2003), as touched on at the beginning of the introduction. In the twenty-first century our situation is far more dangerous than in the 1970s when Schumacher developed his analysis. For example, Rees states that a small group or even a single individual could learn to concoct a biological cocktail deadly enough to kill many millions. This is in addition to ever-increasing threats from nuclear and chemical weapons. And this does not take into account the arguments I have advanced for increasing invisible problems, such as anomie and alienation. Every year the continuing development of technologies for death and destruction brings us ever closer to Armageddon. For example, in January 2007, a group of prominent physicists associated with the *Bulletin of the Atomic Scientists* pushed the hands of its symbolic Doomsday Clock forward from 11:53 to 11:55, five minutes to midnight. In addition to nuclear standoffs with North Korea and Iran and worldwide terrorism, the group saw global warming as a threat to human civilization. All the while there remains almost no awareness that our metaphysical stance coupled with our distorted approach to the scientific method propels us on our path toward oblivion.

Yet Figure I-3 suggests a way out, granting that we have delayed in taking Schumacher's advice for far too long and, as a result, our prospects are not as good as they were in the 1970s. We can indeed work toward changing our worldview—and our scientific method as well—from its stratified direction to an evolutionary direction. As human beings, we have infinite potential to learn from our failures as well as our successes. We can learn, for example, to build on what Schumacher called "the whole cosmos of human thought." We need no longer continue with the "logic of production" in which "the production of deadly devices continues to be deemed a legitimate employment of man's creative powers." Instead, we can turn our attention to producing whatever helps us develop further as human beings. As a result we can learn to not only change our institutions—like educational institutions—but also narrow our aspirations-fulfillment gap, reduce our anomie, and address our fundamental problems ever more effectively. The next section on cultural value fulfillment offers some further hints as to how we might succeed in accomplishing this.

Cultural Value Fulfillment

Let us return once again to Figure 6-1, for it is this "extraordinary language," following Gouldner's assessment (1972), that can help us to see more clearly both our problems and directions for solving them. And let us not forget that language is the human being's most powerful tool for confronting problems ever more effectively. The preceding section focused on anomie and its relationship to the individual's aspirations-fulfillment gap. Let us now look at the concept "cultural worldview" or metaphysical stance. Just as *anomie* refers to the substantial aspirations-fulfillment gaps of individuals widely shared throughout society, so does *cultural worldview* refer to the worldview of individuals widely shared throughout society, for the individual is shaped by social structures, an idea that sociologists and anthropologists emphasize with good reason. We don't come up with the idea of a stratified worldview all by ourselves: we learn those fundamental assumptions as a result of the full range of our experiences in society. Of course, each of us has unique experiences as well, so that we differ from one another to some extent in the nature of our fundamental assumptions. Just as in the case of anomie, different subgroups within society with different subcultures also differ from one another to some extent in their worldview. Yet granting this, our cultural worldview prevails to a great extent in influencing every single one of us.

Following the argument of the anthropologist Donald Brown, let us not make the mistake of emphasizing our differences while avoiding our similarities, such as the common worldview or metaphysical stance within a given culture. Still further—again following Brown's argument—let us look to what contemporary cultures generally have in common, granting the differences among them. It can be useful to emphasize this because of the enormous impact of the scientific and technological revolutions over the past five centuries, based on an initial understanding of the nature of the scientific method. For those revolutions have succeeded in shaping contemporary societies so as to create uniformities across cultures that previously differed from one another to a much greater extent. This by no means suggests that this illustrates human progress, if we take into account the warnings of Schumacher and Rees. The one-sided materialistic development of the modern world has yielded escalating problems that threaten our very survival. Yet at the same time we have developed cultural values—like equality and the worth of every single individual—that point us toward our further sociocultural and psychological evolution. We can see both developments in Figure I-1: the top curve of ever-increasing material and nonmaterial aspirations, and the escalating aspirations-fulfillment gap or anomic situation throughout modern society.

What, then, would be the nature of a cultural worldview that is evolutionary versus stratified? Given the complexity and largely invisible nature of a worldview, the question poses an exceedingly difficult problem. I made a stab at this question—with the help of my colleague, Louis Johnston—in *The Invisible Crisis of Contemporary Society: Reconstructing Sociology's Fundamental Assumptions* (2007). Actually, we took up not just the basic assumptions of sociologists but also those of contemporary society. We attempted, given our substantial personal limitations, to take into account what Schumacher called "the whole cosmos of human thought." Of course, we were able to actually do no such thing, but at least we started on a journey pointing in that direction, and this book illustrates my own continuation on that journey. The part and chapter titles in the contents illustrate the range of phenomena that must be taken into account for any effort to address our cultural worldview: physical structures, personality structures, social structures, and the momentary scene, including long-term history as well. The contents also emphasize the contrast between a stratified and an evolutionary worldview within each of the chapter titles.

If indeed we are to understand the nature of worldviews or metaphysical stances more fully, it is essential that we continue to learn about the extraordinary language of social science, as illustrated in Figure 6-1. In particular, the concept of "cultural values"—the values, interests, ideals, or goals widely shared throughout society—is central to understanding our fundamental assumptions or worldview as well as the nature of culture. Granting the differences of peoples across diverse cultures, I have argued earlier—following Brown's analysis—for the importance of similarities. I have also argued that our continuing scientific-technological revolutions have emphasized those similarities across the globe. Specific phenomena associated with those revolutions—such as the mass media of communication, the computer, and an emphasis on one language (English)—have contributed to those similarities. We can then examine the cultural values that are most closely associated with those revolutions and, as a result, encompass the cultural values central to the great mass of humanity at this time in history. Let us make sure that we understand cultural values as occupying the upper curve of Figure I-1—the curve of aspirations—and not the lower curve of actual fulfillment. Some might argue that aspirations are meaningless without fulfillment and that the humanistic ideals of someone who commits murder are worthless. My argument, instead, is that those ideals are real: they are structures. But the human being is a complex entity, and other values and other structures can—in a given situation—get in the way of ideals. Social scientists have the ideal of following the scientific method, yet certain social structures—such as a stratified cultural worldview—interfere with that ideal. Nevertheless, it is most useful to understand the existence of that ideal, since we might learn to alter those structures that get in its way and thus achieve its fulfillment.

The sociologist widely known for working with the concept of cultural values is Robin M. Williams, Jr. Although his focus was on the cultural values within American society, we might use much of his analysis to indicate the nature of cultural values throughout contemporary societies as a whole. For in modern times the United States has been the most influential nation in the continuing development of the scientific and industrial revolutions, granting the importance of forces throughout Europe, Asia, and elsewhere for establishing the basis for those revolutions. Those forces include, for example, the developments within Greek and Roman civilizations, Judeo-Christian ideals, the invention of the printing press and the creation of the university, Islamic intellectual developments linked to Greco-Roman civilizations, the Renaissance and Reformation in Europe, the journeys of exploration throughout the world, the French eighteenth-century Enlightenment era, and much more. As for the nature of cultural values, Williams includes the following eight that I have divided into two sets: people-oriented cultural values and work-related cultural values. The former are associated most closely with the values emphasized by the French Enlightenment and French political ideals, and also by the American Revolution along with American ideals. These values focus on the political institution. The latter are associated most closely with the scientific and technological revolutions. Their focus, by contrast, is on the economic institution or the world of work. We might see all of these cultural values as specifying the nature of the upper curve of Figure I-1. They are important in themselves, granting that we cannot use their existence to determine the degree to which such values are in fact fulfilled, as indicated by the lower curve of Figure I-1. Here are some excerpts from Williams's analysis:

People-Oriented Cultural Values

Equality. At the level of explicit doctrine, intrinsic equality is widespread in American culture ... in the form of a specifically religious conception (the equality of souls before God.... At the level of overt interpersonal relations, adherence to a sense of intrinsic human value is discernible ... by an extraordinary informality, directness, and lack of status consciousness in person-to-person contacts.... A second major type of equality consists of specific formal rights and obligations ... from military service to voting, from public education to taxation—representing not only freedom but also equality.

Freedom. Always the demand was for freedom from some existing restraint ... a tendency to think of rights rather than duties ... a distrust of central government.... American spokesmen emphasize freedom of speech and assembly, a multiparty, representative political system, private enterprise, freedom to change residence and employment ... it has seemed to make a great difference whether the individual receives a certain income or has a certain type of occupation as a result of an apparently impersonal, anonymous, diffuse, competitive process, as against "being forced" to accept that employment or remuneration by law or by the command of a visible social authority.

Democracy. Along with majority rule, representative institutions, and the rejection of ... monarchical and aristocratic principles ... American democracy stressed the reservation of certain "inalienable rights" as unalterable by majority rule.... The new system was devised in such a way as to limit and check centralized governmental power and to establish an ordered pattern for agreeing to disagree.... Its ... fundamental assumption is the worth and dignity and creative capacity of the individual, so that the chief aim of government is the maximum of individual self-direction, the chief means to that end the minimum of compulsion by the state.

Individual Personality. .... we note a large number of important legal provisions [for] ... the protection of personal freedom or the physical or social integrity of the person ... illegality of slavery ... illegality of imprisonment for debt ... prohibitions against personal defamation (libel and slander); prohibition of "improper search and seizure"; prohibition of "cruel and unusual punishment"; right of habeas corpus.... The "value of individual personality" as impressionistically conceived represents ... uniqueness, self-direction, autonomy of choice, self-regulation, emotional independence, spontaneity, privacy, respect for other persons, defense of the self, and many others.

Work-Related Cultural Values

Achievement and Success. American culture is marked by a central stress upon personal achievement, especially secular occupational achievement. The "success story" and the respect accorded to the self-made man are distinctly American, if anything is. Our society has been highly competitive—a society in which ascribed status in the form of fixed, hereditary social stratification has been minimized. It has endorsed Horatio Alger and has glorified the rail splitter who becomes president.

Progress. By the late nineteenth century ... [p]rogress could now become a slogan to defend the course of technological innovation and economic rationalization and concentration.... Progress became identified with "free private enterprise" ... a belief in the positive value of ever-increasing quantities of goods and services, as illustrated in the dogma that ever increasing per capita gross national product is the touchstone of progress toward "abundance."

Material Comfort. .... a certain kind of materialism may emerge in a society, even though it is not initially a primary criterion of desirability—in the sense

> that the sheer availability of creature comforts and the incessant advertising used to sell them creates a social pressure to concentrate effort and attention upon them ... the objective opportunity to secure material comforts elicits, in the long run, a desire for them. Once a high standard of living has been enjoyed ... it is extremely difficult to reduce the level of sensation.
>
> *Science and Secular Rationality.* Applied science is highly esteemed as a tool for controlling nature. Significant here is the interest in order, control, and calculability—the passion for an engineering civilization.... But the prime quality of "science" is not in its applications but in its basic method of approaching problems—a way of thought and a set of procedures for interpreting experience.... Science is disciplined, rational, functional, active; it requires systematic diligence and honesty; it is congruent with the "means" emphasis of the culture—the focus of interest upon pragmatism and efficiency and the tendency to minimize absolutes.... (Williams, 1970: 454–458, 468–484, 487–489, 492–498)

We previewed these cultural values in chapter 3, where we examined Karen Horney's *The Neurotic Personality of Our Time* (1937). She cites a series of contradictions among cultural values, such as "between competition and success on the one hand, and brotherly love and humility on the other." This contradiction is illustrated above by the values of "achievement and success" and "material comfort" on the one hand, and of equality (the Judeo-Christian ideal of the equality of all souls before God) and individual personality on the other hand. More generally, we have the contradictions between work-related cultural values and people-oriented cultural values.

Horney saw no way out of these contradictions "within the normal range" other than either "to take one of these strivings seriously and discard the other; or to take both seriously with the result that the individual is seriously inhibited in both directions"(1937: 287). But a third alternative was advanced in *The Invisible Crisis of Contemporary Society:*

> A third alternative, one which is outside of "the normal range" ... is to reject our stratified worldview and point toward an interactive [or evolutionary] worldview, which would also point toward dramatic changes in personality and social structures. Metaphorically, we can understand such a change as movement from a see-saw world to a stairway world, where our own success need not threaten others but rather might help them learn how to achieve their own success. (Phillips and Johnston, 2007: 93)

It is our stratified worldview that appears to be the source of these contradictions between our fundamental cultural values. And a change of that worldview to an evolutionary worldview, which certainly is outside "the normal range" of people's ideas, promises to resolve those contradictions.

More specifically, such a change would involve a broadening of the cultural value of "science and secular rationality, where science would be defined not merely by biophysical science but would include social science as well. As a result, that cultural value would no longer encourage our emphasis on "an engineering civilization," on "material comfort," and on a materialistic civilization. "Achievement and success" would be broadened far beyond our present-day narrow view of occupational achievement as the major direction for advancing humanity. By contrast, our people-oriented values would become ever more important, assuming that a broader approach to science would help us all work more effectively toward fulfilling those cultural values. Further, our overriding emphasis on the political sphere of life could be broadened if indeed we extended our approach to social science into practicing the scientific method in our everyday lives. It is one thing to have a democratic form of government, but it would be quite another thing to see democracy practiced within all of our institutions, including the worlds of work, education, religion, and the family.

Up to now little has been said about the concept of "institutions" that appears among the social structures depicted in Figure 6-1. We might define an institution as a system of cultural values and norms within a society focused on solving a given problem throughout society. Thus, for example, the institution of science centers on the problem of developing understanding of the universe, including physical, biological, social, and personality structures. The institution of education centers on the problem of communicating that understanding to all of us. The economic institution focuses on the problem of producing, distributing, and consuming whatever we come to value. The political institution centers on the problem of the development, distribution, and use of power. The religious institution focuses on the problem of our finding ultimate meaning or significance in life. And the family centers on the problems of both achieving intimate or very close relationships and continuing life from one generation to another.

From this problem-oriented approach, it appears that all of our institutions are faced with a deepening crisis, given increasing threats to the continuing existence of humanity. Apparently, science and education are failing to equip us with the tools we need to solve our problems. Worse, they are succeeding in creating an ever more dangerous world. As for religion, it appears to be more difficult to find meaning in a world with greater threats to human survival. With respect to the family, we are beset with more alienation, addiction, and anomie, all of which work against our ability to achieve close personal relationships with others, as illustrated by rising divorce rates and smaller families. The continuation and even increase of social stratification throughout the world—as illustrated by the growing gap between the rich and the poor—testifies to the political institution's failure to distribute power in accordance with our people-oriented values. Perhaps even more important, there is a failure to develop power, a problem to be discussed in chapter 8, using Gandhi's social movement in colonial India as an illustration. Following Schumacher's analysis, our economic institutions fail in helping us fulfill those people-oriented values to a significant extent. Overall, then, all of our institutions reflect the growing aspirations-fulfillment gap—or the anomie—depicted by Figure I-1. They are all failing us in fundamental ways.

The failure of all of our institutions to cope with the very problems that they were instituted to confront can be seen most easily in the limited range of problems addressed by each institution. For example, in the United States there is presently a focus on the war in Iraq and far less with the basis for the movement of the Doomsday Clock closer to midnight: nuclear proliferation in Iran and North Korea, worldwide terrorism, and global warming. Needless to say, there is almost no awareness of—much less any concern for—the relatively invisible problems depicted by Figures I-1 and I-2. As for the educational institution—including the mass media—Schumacher's critique that "education cannot help us as long as it accords no place to metaphysics" is true now even more than in the 1970s. For example, we have only to look to the daily newspapers, television "news" programs, or newsletters put out by social science disciplines like sociology to become appalled by the narrow, short-sighted topics that avoid the big picture of the problems of the human race. We read about miniscule problems or miniscule solutions—like financial links between drugmakers and prescribing doctors or providing habitats for crocodiles in Florida, granting that such problems should not be ignored. All the while the big problems and efforts to discover big solutions are largely ignored. Whether or not the Emperor Nero actually fiddled while Rome burned, we are doing much the same by using techniques of particularization to narrow our focus on problems and solutions. We are also using the falsification of memory to forget about the Enlightenment and sociological dreams of confronting and solving the problems threatening the continuing existence of humanity. In other words, we are adopting a narrow approach in space and time to our problems—a Flatland approach, as explained in chapter 1—rather than a Spaceland or Timeland orientation.

Yet all of these institutions have been constructed by human beings, and they can be changed by human beings, granting the enormous difficulties involved. Following Figure I-3, a broader approach to the scientific method that we can learn to use in our everyday lives holds that promise for us. In particular, that approach can help us alter our worldview and, as a result, all of our institutions to point us in an evolutionary direction with respect to social and personality structures. In chapter 7 we shall focus not on the "heart," as in this chapter, but on the "head," examining more fully the nature of that broader approach to the scientific method. Indeed, that approach has been the focus of the Sociological Imagination Group over the past eight years. We have emerged with what we call the Web and Part/Whole Approach to the scientific method and have developed a number of monographs (www.sociological-imagination.org). Nevertheless, the group has only begun to open a door to understanding the nature of our worldview or metaphysical stance along with an alternative stance. In chapter 7, however, we shall attempt to move through that door as far as we can to learn how we can change from a stratified to an evolutionary worldview and, as a result, confront more effectively the increasing threats to the very survival of the human race.

CHAPTER 7

"Head"

Bureaucratic Versus Scientific Worldview

I HAVE CONTRASTED A "STRATIFIED" WITH AN "EVOLUTIONARY" WORLDVIEW or metaphysical stance in previous chapters. Yet we can obtain greater understanding of the complex nature of these fundamental assumptions about the nature of the world by using other adjectives as well. In *Beyond Sociology's Tower of Babel: Reconstructing the Scientific Method* (2001), I used the adjective "bureaucratic" instead of "stratified," an adjective that reflects the importance of the concept of bureaucracy. We may note that concept in Figure 6-1, located on the lower right side of the diagram among the concepts relating to "hand," or patterns of action like "social stratification," "social rituals," and "social organization." Although our focus will be on such patterns of action in chapter 8, the concept of bureaucracy can help us here in our effort to understand the contrast between a stratified and an evolutionary worldview. We can also see that contrast as one between a bureaucratic and a scientific worldview.

We may understand a bureaucracy as a group with an extensive hierarchy and division of labor that has been deliberately constructed to solve specific problems. As such, bureaucracies are found throughout our contemporary economic, educational, religious, and political institutions, each of which encompasses a system of cultural values and norms focused on solving a given problem. Bureaucracies thus bring together "heart" (values) and "head" (norms) with "hand" (patterns of interaction and action within a group). By contrast, although families are deliberately constructed to solve such problems as the survival of the human race, they generally have a relatively simple—rather than extensive—hierarchy and division of labor. Max Weber, one of the founders of the discipline of sociology, is credited with analyzing the fundamental features of modern bureaucracies (1964). These include a set of norms or rules that give it continuity, rules that are generally recorded and, thus, depend on the invention of writing within modern society. Those rules create a division of labor within an organization, where each role or office—occupied by an individual or individuals—is recognized as having authority over certain decisions within a specialized area. And those rules also follow the principal of hierarchy, where each lower role or office is under the supervision of a higher one.

With this understanding, bureaucracies are necessarily groups with patterns of social stratification or persisting hierarchy, although individuals with roles or offices within a bureaucracy may advance to higher levels. In addition, bureaucracies have an extensive division of labor. As a result of these two fundamental characteristics of bureaucracies, problems of effective communication develop up and down the hierarchy as well as across the division of labor. Weber saw modern bureaucracies as illustrating a change from the generally irrational nature of traditional organizations to the rationality seen in contemporary societies. Yet severe problems within modern bureaucracies are well illustrated by the Sections within the ASA. It has limited communication not only across its 46 Sections but also across its 300 or so highly specialized subsections. If we assume that the complexity of human behavior demands that sociologists take into account more and more of this knowledge, what results is a gross failure to move toward the organization's goal of advancing our understanding of human behavior. Further, the patterns of social stratification within this organization—while generally less pronounced than in most business organizations—nevertheless interfere with the scientific ideal of openness to all research. Thus, substantially less attention generally is paid to the research of individuals at the lower end of the hierarchy.

We might think of bureaucracies—as illustrated by the ASA—as having developed a double stratification system: one that is vertical and relatively obvious, and another that is horizontal and relatively hidden. We have already discussed the obvious one of hierarchy, with individuals at the upper reaches given more power and privileges than those at the lower reaches. But horizontal stratification is far less obvious. It is based on a detailed division of labor, such as the division of sociology by the ASA into no fewer than 46 Sections, with limited communication among individuals in the various Sections. That limited communication appears to be based on the belief that work in one's own Section is complete unto itself, and other Sections have little knowledge of genuine importance to one's own work. One is polite to those in other Sections, just as individuals in contemporary society generally are polite on the surface to those of other races. Yet beneath that politeness is an indifference and often deep prejudice. I recall walking for many years the corridors of the Department of Sociology at Boston University. I would pass others in different specialized areas like ships in the night, sometimes stopping for a few moments to make small talk. Yet I and others felt that we had little to say about the truly important problems addressed by sociology. The annual meetings of the ASA also reflect this double stratification.

Generally, sociologists take seriously only the papers presented in their own Sections or, in some cases, more than one Section. Because of literally hundreds of sessions and papers, there is little time to learn what is going on within one's own Section, let alone the work in others.

There has been a large amount of research on the workings of bureaucracies, much of which contradicts Weber's belief in their rationality and effectiveness. One study in particular has challenged Weber's belief by contrasting a "bureaucratic" pattern of organization with a rational or scientific pattern (Udy, 1959). Stanley Udy based his research on the work of Helen Constas (1958), who found that key elements within Weber's definition of bureaucracy were not functional or effective for the development of rationality in organizations. On the basis of a variety of Weber's writings, Udy listed seven characteristics of bureaucracy that Weber specified, with the first two corresponding to the division of labor and extensive hierarchy (with three or more levels) cited previously. The other five characteristics were: (3) rewards distributed according to one's position in the hierarchy of positions or offices; (4) limited objectives so as to enable the organization to achieve a focus; (5) an emphasis on performance so that rewards are dependent on achievement; (6) specification of limitations on the involvement of those in the organization, such as with contracts; and (7) rewards given to those with lower authority in return for their participation. Following the work of Constas, Udy saw the first three characteristics of Weber's definition as "bureaucratic" and the last four as rational or scientific.

Selecting 150 organizations producing material goods from 150 societies, Udy emerged with findings that supported the work of Constas. He found that the three bureaucratic characteristics were interrelated, as were the four rational characteristics, and that the three bureaucratic characteristics tended to occur when the four rational characteristics did not, and vice versa. He concluded that grouping all seven characteristics into one definition distorts the opposition between bureaucracy and rationality. If we assume, as sociologists generally do, that Weber's definition does indeed reflect the actual nature of bureaucracies, then the work of Constas and Udy suggests the existence of fundamental contradictions within bureaucratic organizations throughout the world. And those contradictions reflect what research generally has uncovered about such organizations. Far from the rationality and effectiveness that Weber saw, taking into account his focus on their advance over organizations in traditional or preindustrial societies, bureaucratic organizations have serious difficulties in solving the problems they have been constructed to solve. This suggests that the worldview or metaphysical stance guiding their construction has serious limitations that have yielded the contradictions that Constas and Udy discovered between the bureaucratic and rational or scientific elements of bureaucratic organizations.

My purpose here is by no means to summarize existing research on bureaucracies, a task that would extend far beyond the scope of this book.

Instead, my focus is on a topic that has been almost completely neglected by that research literature: the metaphysical stance or worldview that lies beneath and shapes the nature of these organizations. For example, if we see bureaucracies as having a double stratification system—based on its hierarchies as well as a division of labor that leads to more subtle patterns of stratification—then we can indeed understand that they illustrate a stratified worldview. If, as sociologists have discovered, bureaucratic organizations are prevalent throughout the institutions of contemporary societies, then we can well understand the prevalence of a stratified or bureaucratic worldview or metaphysical stance. In the first section of this chapter—"Bureaucratic Worldview"—I will use one of ASA's publications to illustrate the departure of this bureaucracy from scientific ideals. In the second and final section—"Scientific Worldview"—I will suggest directions that ASA might adopt to follow scientific ideals. In both sections I will invoke the fundamental assumptions, worldviews, or metaphysical stances that are involved. I will also include material not specifically linked to bureaucratic organizations, yet oriented to clarifying the contrast between a bureaucratic or stratified worldview and a scientific or evolutionary one.

Bureaucratic Worldview

My focus is on the February 2007 issue of *Footnotes,* a monthly newsletter sent to all ASA members. I make no claim that this issue is representative of the newsletter, although my own personal sense is that it is typical. My purpose is to examine the degree to which this publication follows the ideals of the scientific method and also the extent to which it illustrates a bureaucratic (or stratified) versus a scientific (or evolutionary) worldview. To achieve even greater focus, I concentrate on one scientific ideal in particular: pointing in a reflexive direction to open up to the possibility of assessing and improving the research process itself. Such a reflexive direction is fundamental to an evolutionary or scientific worldview, as discussed in the preceding chapters, and is illustrated by my own analysis of the failings of the ASA. Given the failure of the social sciences to achieve much in the way of rapid cumulative development and their general failure to understand and confront effectively basic problems in society, such a reflexive orientation is both important and timely for the further development of the scientific method as applied to human behavior.

Looking to this issue of *Footnotes,* a reflexive orientation in its pages would point toward the problems experienced by sociologists in general as well as by individual sociologists. For example, there is the problem of "investigator effect": to what degree does the behavior of investigators—such as their basic assumptions—work to influence the behavior of those whom they are investigating? There are also the problems suggested by Constas and Udy in their analysis of bureaucratic organizations. For example, to what degree does a division of labor—with 46 distinct Sections of the ASA—yield severe problems among sociologists and reduce the possibilities for living up to scientific ideals? To what degree do patterns of hierarchy throughout the discipline of sociology also militate against scientific ideals? Do the materials encourage, directly or indirectly, limited communication among sociologists in different specialized areas, or do they encourage the reverse? Do they emphasize hierarchies among sociologists? Do they encourage sociologists to examine their own basic assumptions? Their research procedures? Their impact on others outside the discipline? Do they ignore the range of issues relating to reflexive behavior?

Overall, all the articles and other materials in the sixteen-page issue of *Footnotes* except one avoid any focus on the problems of sociology or of individual sociologists, choosing instead to emphasize the achievements of the discipline and of individual sociologists. There is no question that there are definite achievements. They are not the kinds of achievements, however, that point toward the importance of making fundamental changes in the discipline as a whole or by individual sociologists, changes such as questioning the nature of the metaphysical stance or worldview held by sociologists. For example, *Footnotes* has an article on Ricardo Lagos, former president of Chile, who succeeded in shaping that country's democratic orientation and who was invited to speak at ASA's 2007 meetings in New York. There are articles on the many achievements of sociologist Seymour Martin Lipset who died in 2006 and on the history of New York's immigration experiences. ASA's executive officer reports on new efforts to communicate the doings of sociologists in mass media outlets, and information elsewhere discusses the successes of sociologists in achieving such communication. Several articles detail awards—presented by ASA as well as by other organizations—given to sociologists as well as undergraduate sociology students. ASA's task force gives its recommendations on how the discipline can contribute to general education. One article describes the background of two sociologists who are ambassadors, one the U.S. ambassador to Trinidad and Tobago and the other Afghanistan's ambassador to Germany. The newsletter profiles the work of a retired sociologist who is helping rebuild New Orleans. There are the usual announcements about meetings, funding, competitions, new books, new programs and deaths, and calls for papers to be presented at conferences.

There is absolutely no question that these articles are all worthwhile. Generally, they point toward the fulfillment of basic cultural values, such as those examined in chapter 6: equality, democracy, individual personality, achievement and success, science and secular rationality. Yet they are not oriented in a reflexive direction, seeking to understand the limitations of the discipline and how to confront those limitations effectively. More specifically, they do not point out to sociologists the importance of examining their own fundamental assumptions or metaphysical stance as a means to learni about the nature of those limitations. The retired sociologist—Thomas

Lief—points out this lack indirectly: "I really believe sociologists have not yet understood society; they research around it, but have not yet reached a meaningful comprehension." That statement is an example of a reflexive orientation. An article in *Footnotes* by John Angle, "A Mathematical Sociologist's Tribute to Comte: Sociology as Science," also adopts such an orientation. Its focus is on Angle's achievements in developing mathematical models of human behavior, with the implication that sociology must move in this direction to become genuinely scientific. Granting the future potential of such models, in my own view they are a narrow approach to understanding the complexity of human behavior. Angle should be commended, however, for raising the issue of sociology's future.

As I read this issue of *Footnotes,* I cannot help but think of what has happened to the Doomsday Clock managed by the group of physical scientists who publish *The Bulletin of Atomic Scientists,* as mentioned in chapter 6. They have moved the hands of the clock from 11:53 to 11:55, based largely on fears of nuclear devastation as well as uncontrollable and severe climate change. Even granting the efforts of sociologists, other social scientists, and millions throughout the world to work toward solving the fundamental problems of human beings at this time in history, we appear to be moving ever closer to midnight.

As I write this, I've just received an e-mail from the ASA, which reads in part:

> **2007 Member-Get-A-Member Campaign**
>
> The American Sociological Association challenges ASA members to encourage their colleagues, students, and others interested in the discipline of sociology to join ASA for 2007. Each current member who sponsors a new member is eligible for special recognition and 2007 rewards.
>
> In our 2006 campaign, more than 100 members brought at least one new member into the Association and received discounts on their 2007 dues. For this year's campaign, sponsors are eligible for discounted 2008 membership dues and our grand prize—a $500 Amazon.com gift certificate.

Here we see what is most important to bureaucracies: their continuation and expansion. Where is the concern over what is wrong with the organization and with sociology as a discipline? Where is the understanding of the problems faced by the human race, paralleling the concerns of the physical scientists who operate the Doomsday Clock? Shouldn't sociologists, who should be aware of the immensity of modern problems, be working along with those physical scientists instead of offering rewards for bringing in new members?

Yet it is unfair to single out social scientists to blame for the failures of contemporary societies in addressing escalating social problems effectively;

we are all in this together. Physical scientists generally have avoided efforts to help push for the development of the social sciences, adopting instead the relatively easy path of understanding much simpler phenomena. Educational leaders, such as university presidents, generally have not understood that we human beings are in a race between understanding our complex problems and our own annihilation, nor have they understood the role of social scientists in helping us win that race. Political leaders have not given social scientists the encouragement and funds to develop a crash program for integrating knowledge of our complex, threatening, and escalating problems. Indeed, we might see that all institutions of contemporary society, with their specialized concerns for solving diverse problems, are nothing less than a huge bureaucracy that has organized society as a whole. Many patterns of social stratification exist throughout society, as illustrated by the growing gap between the rich and the poor. There is also the more subtle kind of stratification associated with the division of labor among institutions, as illustrated by the powerful and advantaged situation of the political and economic institutions by contrast with the educational institution. From this perspective we can understand—although we need not condone—the situation of the social sciences in contemporary society. The social sciences are victims of a bureaucratic society that functions to leave them no more than the crumbs of its funds, diverting them instead, for example, to the military or to the physical sciences that can aid in the development of a materialistic economy.

We might contrast the present state of sociology with its situation in the middle of the twentieth century. For example, there appears to have been great optimism among sociologists for fulfilling what Mills called "the promise of sociology," the possibility of helping modern society solve its fundamental problems. Mills revealed his own optimism—perhaps a central reason that he continues to be revered by sociologists despite their own growing pessimism and cynicism about the future of their discipline—in *The Sociological Imagination*:

> it is by means of the sociological imagination that men now hope to grasp what is going on in the world, and to understand what is happening to themselves as minute points in the intersections of biography and history within society.... The sociological imagination is the most fruitful form of this self-consciousness. By its use men whose mentalities have swept only a series of limited orbits often come to feel as if suddenly awakened in a house with which they had only supposed themselves to be familiar. Correctly or incorrectly, they often come to feel that they can now provide themselves with adequate summations, cohesive assessments, comprehensive orientations. Older decisions that once appeared sound now seem to them products of a mind unaccountably dense. Their capacity for astonishment is made lively again. They acquire a new way of thinking, they experience a transvaluation of values: in a word, by their

> reflection and by their sensibility, they realize the cultural meaning of the social sciences. (1959: 7–8)

Up to this point in my examination of our bureaucratic or stratified worldview I have focused on the discipline of sociology. Yet a cultural worldview can extend very widely, far beyond the academic world, granting the impact of that world on society as a whole. The pessimism we see within academia is reflected outside academia. One among many examples of this pessimism is the contemporary emphasis on disaster films, portraying such phenomena as nuclear fallout, sinking ships, alien attacks, avalanches, earthquakes, floods, meteors, vicious twisters, and killer bees. Films such as *Titanic, The Towering Inferno, Kingdom of the Spiders,* and *Armageddon* appear to reflect the mass slaughters we have experienced in the twentieth and twenty-first centuries along with our failure to learn how to escape from those experiences. Such films are more than merely escapist entertainment. They teach us that we have no way out of a future where the continuity of the human race is threatened, avoiding any optimistic lessons about our ability to cope with such threats.

My orientation is by no means to spread a doom-and-gloom philosophy, but rather to alert us all to increasing dangers so that we can learn to confront them effectively. This contrasts with what we generally do at present: largely bury the dangers from our awareness, just as the Springdalers engaged in the technique of particularization and the falsification of memory (Vidich and Bensman,1960), and just as sociologists do when they specialize, have limited communication outside their specialties, and forget about "the promise of sociology." We might think here of a pendulum as a metaphor or image of the way that the scientific method should work, a pendulum that swings in ever-widening arcs. In one direction it swings toward deeper awareness of a problem, such as increasingly deadly biophysical technologies. In the other direction it swings toward understanding a problem and creating the basis for confronting it effectively. Such a metaphor does justice to the first and most important step of the scientific method: the definition of a problem. It helps the individual using the scientific method—whether a professional scientist or any one of us—bring to bear his or her emotions on the situation at hand. By so doing it helps us become deeply committed to learning how to use the scientific method in an ever more effective way.

Continuing to explore the nature of a bureaucratic or stratified worldview by looking outside the academic world, we might turn to the mass media in general and television in particular for further illustrations. One study described in some detail in *Beyond Sociology's Tower of Babel* (Phillips, 2001: 101–104; see also Hornig, 1990) takes a critical look at the status of the social sciences conveyed by *NOVA,* the award-winning PBS documentary science series. The series attempts to portray the nature of modern science. Yet apparently what it succeeds in accomplishing is depicting the nature of a bureaucratic or stratified approach to the scientific method, an approach where there is little or no room for viewing the social sciences as genuine sciences. Further, there is little or no room for the rest of us to imagine that we might learn to use the scientific method in our everyday lives. For example, in Hornig's analysis of one episode—"The Race for the Superconductor"—the portrayal of superconductors emphasizes their "almost magical powers," repeatedly portraying a cube that appears to be dancing in air. In this way *NOVA*'s producers take the scientific method far away from the viewer's everyday experiences. The physical scientists involved are portrayed as using intelligence far above that of us ordinary mortals, given their blackboards covered with complex equations and their bottles of liquids dripping through complicated arrangements. The drama of the show centers not on the technological and scientific problem of how to produce a superconductor but rather on the competition between groups from Sweden, Japan, and the United States, with most viewers identifying with the United States in its efforts to win the race for the superconductor. Of course, in episode after episode of *NOVA* social scientists are nowhere to be seen.

Moving from documentaries to fiction, Jane Banks and Jonathan David Tankel (1990) have analyzed *Star Trek* and *Star Trek: The Next Generation*, two very popular series that continue to be rerun. Apparently they join *NOVA* in invoking a hymn to the wonders of physical and biological science and their technologies. They succeed not only in devoting little attention to the social sciences but also in emphasizing the glories of our own technological civilization, all the while largely ignoring the fundamental problems within our civilization. The contradictions within our own way of life—such as the increasing aspirations-fulfillment gap depicted in Figure I-1—are well illustrated by Captain Kirk, Spock, and McCoy of *Star Trek*, as well as by Picard, Riker, and Worf of *Star Trek: The Next Generation*. Although the "Prime Directive" guiding the starship *Enterprise* is to avoid interfering with the evolution of life and societies encountered anywhere in the universe—suggesting the overriding importance of the cultural value of equality—the crew of the *Enterprise* succeeds in transforming the cultures they encounter in the direction of the so-called advanced culture of modern society here on Earth. This stratification among the cultures throughout the universe is a far cry from the cultural value of equality.

In analyzing further the nature of our bureaucratic or stratified worldview, we might do well to examine two of our institutions: education and science. Of course, we can do no more here than take a quick glance at them, yet that may nevertheless be of some use. For it is these institutions more than the others that center on what distinguishes us human beings from all other forms of life: our capacity to continue to learn—with the aid of language—without any limit whatsoever. George A. Lundberg published *Can Science Save Us?* in 1947 and revised it thoroughly in 1961. His analysis centers on fundamental problems that these institutions reveal in their efforts to develop and communicate the understanding of human behavior that we humans desperately need in the modern world if we are to survive:

> What are the principal defects in our educational program as it stands? The first defect of our present educational program is fortunately one on which all seem to agree, including even Mr. Hutchins and Mr. Dewey.... I agree with him [Hutchins] that modern education suffers from the lack of a unifying discipline, such as theology provided in the Middle Ages. I contend, however, that this unifying discipline today cannot be, as some feel, an amorphous mass of one hundred or any other stated number of books, called "the classics," dealing primarily with the ancient and medieval gropings of mankind to find its way about. The unifying discipline of modern education lies in modern science, which is simply the distilled essence of the classics of all time. Science is the positive residue of the sifting and winnowing of all the ages, and its methods are the demonstrably effective instruments of reaching valid conclusions....
>
> If we agree that our failure to recognize the dominant importance of the scientific method is our first great oversight, then the second principal defect of our educational program is clearly our failure to develop, even in our college students, an adequate knowledge of the elements of that method ... there is the deep-seated notion ... that science has to do exclusively with the physical world and that its methods are not applicable to human affairs. Why is it that we find ourselves with thousands of research laboratories devoted to the application of the scientific method to physical problems and almost none, comparatively speaking, devoted to research in human relations? Is it because human relations are functioning so smoothly that they need no attention? A glance at industry, our minority problems, and the international situation should be sufficient answer. Is it because, while physical research pays a tangible dividend in profits, improvement in human relations does not? Ask any employer what labor troubles, strikes, lockouts, slowdowns, and loafing is costing him, and in the end, costing all of us. Then why don't we risk a few million dollars in scientific research on human relations the way we risk hundreds of millions on industrial research, because in the end we know it will pay? There can be only one answer: We do not yet believe that human relations is a subject to which the scientific method is applicable.... To keep nations from flying at each other's throats, any political hack will do. Human relations will improve when we undertake serious scientific study of how to improve them ... such is the preoccupation in the teaching of science with specific subject matter and with specific technologies, that many students, although they major in some particular science, have no conception of the scientific method as a generally valid approach to the problems of this world. Science is generally thought of as a type of *subject matter* rather than as a *method* of study.... So superficial is the habit of

> rational and scientific thinking about social events that in situations at all critical (e.g., war), even the scientifically trained promptly abandon this comparatively recent acquisition in favor of a more primitive and emotional personalistic method....
>
> The third great defect of our educational program at present is its failure to equip students with an adequate understanding of the nature and uses of language. Just as it is impossible to teach physics without some understanding of mathematics, so it is impossible to teach any social science without an understanding of the particular kind of language usage and logic which all science absolutely requires. Indeed, the first lessons in scientific methods, in their broader meaning, as contrasted with mere laboratory techniques and scientific information, must be concerned with the nature and uses of language and logic. (Lundberg, 1947/1961: 75–80)

Lundberg's book is hardly known among contemporary sociologists, yet his systematic understanding of the problems of the modern world appears to rival that of C. Wright Mills and Alvin W. Gouldner. His view that the unifying discipline throughout the institution of education should be science points away from our bureaucratic or stratified worldview that sharply separates these two institutions. We might proceed to extend his view to the idea that science might become the unifying discipline throughout all of our institutions and, thus, become the basis for a scientific or evolutionary worldview. Lundberg does not mean, of course, the one-sided kind of science emphasized by *NOVA* and *Star Trek,* but rather the kind of science that emphasizes the overriding importance of social science. From the perspective of this book and the benefit of what has been learned over the past half century, we can extend the breadth of our understanding of science so as to build on what we have learned from Mills's *The Sociological Imagination,* as discussed in the foregoing chapters. That extension would take seriously what Lundberg claims about emphasizing the importance of the scientific method as distinct from the subject matter of the various sciences; the contributions of philosophy, religion, literature, and the arts to our understanding of the scientific method as well as human behavior; and the importance of language for our ability to use the scientific method. Following Gouldner, unless we learn to use the extraordinary language of social science—as illustrated by Figure 6-1—we have severely limited our understanding of human behavior.

Lundberg has much to say about the urgency of moving beyond these kinds of insights to the practical matter of changing the institutions of education and science. But how are we to actually accomplish such vast changes in the modern world? How are we to overcome the strong objections that would arise from within every single institution and from people in general throughout the world? Granting the breadth of Lundberg's approach, we must extend it further from our vantage point a half century later. For example, there is Lundberg's easy dismissal of Robert Hutchins's vision of education as emphasizing an "amorphous mass of one hundred or any other stated number of books, called 'the classics,' dealing primarily with the ancient and medieval gropings of mankind to find its way about." Those books, and a great many others, are exceedingly important for us to understand if we are indeed to learn the nature of human behavior. They can be encompassed with a very broad approach to the scientific method, such as the approach I have attempted to put forward in this book. Given such breadth, we must still confront the problem that is the focus of this book: how to move toward altering our institutions so that we all learn to use the scientific method in our everyday lives? Understanding the nature of a scientific or evolutionary worldview appears to me central to that task. The final section of this chapter focuses less on this problem and more on directions for solving it. The pendulum illustrating the scientific method will thus swing from one side to the other. As it does so, let us realize that our awareness of and commitment to this problem—learning to use the scientific method in everyday life—is the most important basis for developing an understanding of how to solve it.

Scientific Worldview

A fundamental relationship exists between a bureaucratic and a scientific worldview, following the metaphor of the pendulum. Without understanding the problems linked to a bureaucratic or stratified worldview, we will have no motivation to alter it. Conversely, unless we have at least a vague notion of some alternative it—such as a scientific worldview—we will remain unable to confront our bureaucratic worldview. Otherwise, we would be cutting the ground out from under ourselves with no other piece of ground to step on. This point was made clearly by Thomas Kuhn (1962), briefly referred to in the introduction as well as in chapters 1 and 2, and examined at greater length in *Beyond Sociology's Tower of Babel* (Phillips, 2001: 50–55). Kuhn's *The Structure of Scientific Revolutions* (1962) has proved most influential among social scientists and beyond, even though his focus was on the history of science. What he succeeded in doing was to show how important are the fundamental assumptions or the "scientific paradigms" of scientists. For old theories cannot be so easily overthrown by new theories based on different scientific paradigms no matter how much evidence is advanced in their favor. New theories must also be accompanied by contrasting the old and the new scientific paradigms and showing how the latter promise to resolve contradictions or problems within the former. This approach suggests much the same for even broader metaphysical stances or "cultural paradigms." To change our bureaucratic metaphysical stance or paradigm, we must have in hand an alternative paradigm—in our case, a scientific worldview—that promises to resolve the contradictions within a bureaucratic worldview.

Kuhn also alerted us to the enormous difficulties in changing a scientific paradigm. Factors such as tradition, hierarchy, and the personal rigidity of scientists—factors closely linked to a bureaucratic approach to the scientific method—play an important role in blocking such basic changes. For example, the shift from Newton's laws of motion to Einstein's theory of relativity was made by physicists who all their lives had accepted the validity of Newton's laws without question. As a result it took many years before Einstein's theory was accepted. If changes in scientific paradigms have proven difficult, we might well expect changes in cultural paradigms, worldviews, or metaphysical stances to be far more problematic. For these latter assumptions affect every aspect of our behavior from one moment to the next in every situation that we experience. Nevertheless, Kuhn's understanding of scientific revolutions provides us with a guide to understanding how such basic changes might take place. Whatever else is involved, we require an alternative worldview that promises to resolve the contradictions or problems within the original worldview.

Given the extremely abstract or intangible nature of worldviews or metaphysical stances, it is useful to examine the work of others who have indicated their importance, And given the difficulties of changing fundamental assumptions, such examinations can help motivate us to commit to this difficult enterprise. We might turn initially to William James, one of the founders of the philosophy of pragmatism, a philosophy that provides the scientific method with a broad framework of assumptions. James quotes from an essay by G. K. Chesterton, an English essayist, novelist, and poet who wrote in the early part of the twentieth century:

> There are some people—and I am one of them—who think that the most practical and important thing about a man is still his view of the universe. We think that for a landlady considering a lodger, it is important to know his income, but still more important to know his philosophy. We think that for a general to fight an enemy, it is important to know the enemy's numbers, but still more important to know the enemy's philosophy. (James, 1907/1995: 1)

James has in mind philosophy in general and also metaphysics in particular. For he goes on to make the claim that, within the new philosophy of pragmatism, "science and metaphysics would come much nearer together, would in fact work absolutely hand in hand" (1907/1995: 20). His argument is not simply that metaphysics is an important area of knowledge for scientists. He finds an understanding of metaphysics to be crucial for all of us in our everyday lives. In addition, he sees one's metaphysical stance as shaping all behavior. James's belief is much the same as my own with reference to this book. Granting the complexity and almost invisible nature of cultural paradigms, worldviews, or metaphysical stances, there are no beliefs of individuals that are more important, for they shape all behavior. If indeed modern society is experiencing threats to its very survival, then—whatever else we examine—we must turn to the nature of our worldviews if we wish to confront those problems effectively.

There have been references to the ideas of Gouldner—with his focus on the importance of the extraordinary language of social science—in the previous chapters, especially in the introduction and in chapter 1. More discussion of some of his key ideas appears in *The Invisible Crisis of Contemporary Society* (Phillips and Johnston, 2007: 52-57). He comments on the importance of metaphysics or what he calls "background assumptions":

> Background assumptions ... influence the social career of a theory, influencing the responses of those to whom it is communicated. For, in some part, theories are accepted or rejected because of the background assumptions embedded in them.... Background assumptions come in different sizes, they govern domains of different scope. They are arranged, one might say, like an inverted cone, standing on its point. At the top are background assumptions with the largest circumference, those that have no limited domain to which alone they apply. These are beliefs about the world that are so general that they may, in principle, be applied to any subject matter without restriction. They are, as Stephen Pepper calls them, "world hypotheses." Being primitive presuppositions about the world and everything in it, they serve to provide the most general of orientations, which enable unfamiliar experiences to be made meaningful. They provide the terms of reference by which the less general assumptions, further down the cone, are themselves limited and influenced. World hypotheses are the most pervasive and primitive beliefs about what is real.... World hypotheses—the cat may as well be let out of the bag—are what are sometimes called "metaphysics." (Gouldner, 1970: 29-31)

Just as Kuhn recognized the importance of scientific paradigms for physical scientists, so does Gouldner—a sociological theorist—recognize the centrality of background assumptions for sociological theory.

The individual who played the most important role in founding the philosophy of pragmatism is Charles Peirce, some of whose contributions to our understanding of the scientific method were discussed in chapter 4. In our examination of metaphysics it is useful to remind ourselves of his position:

> There is and can be no doubt that this immature condition of Metaphysics has very greatly hampered the progress of ... psychology ... linguistics, anthropology, social science, etc. To my mind it is equally clear that defective and bad metaphysics has been almost as injurious to the physical sciences.... The common opinion has been that Metaphysics is backward because it is intrinsically beyond the reach of human cognition.... But metaphysics, even bad metaphysics, really rests on observations, whether consciously or not; and the only reason that this is not universally recognized is that it rests upon kinds of phenomena with which every man's

experience is so saturated that he usually pays no particular attention to them. (1898/1955: 310–311)

Yet it is one thing to discuss the importance of metaphysics in general and quite another to attempt to specify the nature of our own metaphysical stance or of an alternative stance. This is what Ervin Laszlo, a philosopher discussed in chapter 4, attempted to do in his book, *Introduction to Systems Philosophy* (1973). Yet systems philosophy falls in the trap of emphasizing achievements in the physical and biological sciences at the expense of the social sciences. Laszlo ignores pragmatist philosophers such as James and Peirce. Following the arguments of Lundberg, modern society for far too long has treated social science as a stepchild, failing to recognize the enormous complexity of human behavior and failing to give social science the encouragement and funding it desperately needs. If, however, we reverse this trend, then, in Lundberg's words, "Social research institutions will make their appearance, which will rank with Massachusetts and California Institutes of Technology, Mellon Institute, the research laboratories of Bell Telephone, General Electric and General Motors, not to mention several thousand others" (1947/1961: 143).

My previous efforts to examine the nature of our worldview and an alternative one—coupled with those of Louis Johnston—may be seen primarily in *The Invisible Crisis of Contemporary Society* (Phillips and Johnston, 2007). Our overall finding was that a bureaucratic worldview works to shape an increasing aspirations-fulfillment gap, and an interactive, evolutionary, or scientific worldview works to decrease that gap. That relationship appears in Figure I-2, where the increasing gap is linked to increasing personal and social problems. But this summary of the book's major thesis fails to do justice to the many authors whose work appears here and who contributed substantially to an understanding of these two worldviews or metaphysical stances.

Given these general ideas about alternative worldviews, it is useful to return to our focus on the role of organizations in pointing us in the direction of one worldview or the other. In the analysis of bureaucracy by Weber and Udy in the introduction to this chapter, I noted Udy's distinction between Weber's first three characteristics, which were bureaucratic, and the other four that were interrelated and rational or scientific. Let us look more closely at the nature of the latter four characteristics, for they hold clues as to how we might succeed in constructing organizations further away from a bureaucratic worldview and closer to a scientific worldview. In particular, two of them concern rewards given to those working in bureaucratic organizations: (5) an emphasis on performance so that rewards are dependent on achievement, and (7) rewards given to those with lower authority in return for their participation. We should note that such rewards are given out by those at the top of the organization to those lower down, thus reinforcing the organization's hierarchical structure. We should also note that even though it is the intention of an individual giving out rewards to reinforce achievement and participation, in fact such rewards may easily work simply to reinforce the authority of those dispersing them. Nevertheless, the rewards—with their egalitarian orientation—suggest an important direction for bureaucratic organizations that moves them away from a strictly bureaucratic worldview.

We might raise the question, however, of just how reinforcing such rewards might actually prove to be, granted that they will have some impact given the situation throughout society that there is a large and growing aspirations-fulfillment gap associated with our bureaucratic or stratified worldview. Such rewards are designed to help those receiving them fulfill certain aspirations. If many other aspirations remain unfulfilled—as appears to be the case in modern society as a whole—then any given reward must become a relatively limited one and does little to reduce an individual's overall aspirations-fulfillment gap. In other words, such rewards are limited in their effectiveness in motivating individuals to improve their performance and commitment to the organization in question. When we add to this situation that the giving of such rewards solidifies the organization's hierarchical structure, we begin to understand how little bureaucratic organizations can do to move toward a rational or scientific or evolutionary worldview without basic changes in that direction throughout the organizational culture. Yet we should not discount the impact of small changes if they continue to be made over time. Those require, however, a clear vision of an alternative worldview that provides the organization with a direction or some image of the future.

In our analysis of social structures, the focus in this chapter has been on "head" and in the preceding chapter on "heart." What remains is to examine the "hand" or the patterns of action and interaction to be found within social structures. Culture is very important, but so is social organization. Of course, our separation of these three categories is for purposes of mounting a detailed analysis of each one, but we know that in reality they all operate together. For example, it was essential to introduce the concept of bureaucracy as we proceeded to focus on culture. Having done so, we are now in a position to move much further in our analysis of patterns of social organization such as bureaucracy and social stratification. As we proceed to analyze the patterns of social organization of contemporary society, our focus continues to be on questions relating to the relationships sketched in Figure I-2. How does a bureaucratic worldview and scientific method succeed in increasing our aspirations-fulfillment gap along with personal and social problems? As for Figure I-3, how does an evolutionary worldview and scientific method succeed in decreasing that gap and those problems? And how, given such understanding, can we learn to use an evolutionary scientific method in our everyday lives?

Chapter 8

"Hand"

Stratification Versus Conscious Evolution

As we proceed in this chapter to contrast social stratification with conscious evolution, we shall learn more about the importance of the links between social and personality structures that the extraordinary language of social science reveals. For example, social rituals teach us to develop individual rituals emphasizing the trees in our momentary scenes rather than the forest of our narrow worldview that avoids the implications of failing to evolve as human beings. In other words, our patterns of action encouraged by our various institutions shape our individual patterns of action in the situations of our lives, patterns that conform to the limitations of a bureaucratic or stratified worldview. Of course, once we understand that our behavior fosters our large aspirations-fulfillment gap, and once the extraordinary language of social science provides a vision of alternative patterns of action, we can learn to narrow that gap. More specifically, we can learn to develop an East-West strategy of behavior. For example, we can lower our aspirations in the short run, just as Buddha recommended in his advice to reduce our "thirst" and Durkheim recommended when he advised us to calm the "greed" and "anomy" encouraged by the successes of the industrial revolution. As a result of this narrowing of the aspirations-fulfillment gap, the reinforcements that we obtain can give us the encouragement and confidence we require to follow the Western emphasis on the cultural value of progress, only defining such progress to include not only economic development but also the broader evolution of the individual and society.

Stratification

If social stratification is inevitable, then there is no sense in looking for ways to shift from stratification to conscious evolution. There have been debates within the sociological literature on this question of the inevitability of stratification, given the universal existence of such patterns. For example, Marx's theories assumed no such thing, pointing instead to the inevitability of a revolution that would ultimately yield a classless society. For example, Kingsley Davis and Wilbert Moore advanced a theory favoring the inevitability of stratification (1945). They claimed that unequal rewards throughout society's occupational structures are essential to motivate the most qualified individuals to fill the most important positions in society. Dennis Wrong challenged their argument, claiming instead that existing systems of occupational stratification are so detailed that they far exceed any requirements for motivating qualified individuals (1959). My own approach is that the idea of inevitability is almost invariably unscientific, since our understanding of any phenomenon is always limited, and the claim of inevitability assumes the existence of perfect knowledge. For example, this debate fails to take into account the nature of society's worldview. If we have a stratified worldview that can be changed, then patterns of stratification themselves can be changed. The fact that degree of stratification varies from one situation to another suggests the possibility of lessening the degree of stratification in any given situation.

In the introduction to this book I wrote that there is a focus on a different aspect of social stratification within various Sections of the ASA. For example, some specialists do research on sexism, others on classism, others on ageism, still others on ethnocentrism. Granting the importance of such specialized knowledge, specialists fail to emphasize the importance of the general concept of social stratification, for attention to that idea would alert each specialist to the significance of works on other aspects of stratification. Such a specialized orientation to the phenomenon of stratification fails to make progress in understanding the overall significance of social stratification for contemporary society.

Ivan Illich, an educator not involved in any of sociology's specialized areas that have to do with stratification, has come up with such a general understanding, one that invokes the stratification across the various institutions of society:

> Many students, especially those who are poor, intuitively know what the schools do for them. They school them to confuse process and substance. Once these become blurred, a new logic is assumed: the more treatment there is, the better are the results; or, escalation leads to success. The pupil is thereby "schooled" to confuse teaching with learning, grade advancement with education, a diploma with competence, and fluency with the ability to say something new. His imagination is "schooled" to accept service in place of value. Medical treatment is mistaken for health care, social work for the improvement of community life, police protection for safety, the rat race for productive work. Health, learning, dignity, independence, and creative endeavor are defined as little more than the performance of

> the institutions which claim to serve these ends, and their improvement is made to depend on allocating more resources to the management of hospitals, schools, and other agencies in question....
>
> Once a man or woman has accepted the need for school, he or she is easy prey for other institutions. Once young people have allowed their imaginations to be formed by curricular instruction, they are conditioned to institutional planning of every sort. "Instruction" smothers the horizon of their imaginations. (Illich: 1972: 1-2, 55-56)

Whereas most sociological studies of stratification emphasize the occupational system, Illich looks to the range of society's institutions. He uses the concept of "schooling" as a metaphor for stratification. Its advantage is that it indicates the process that is involved in constructing a given hierarchical or unequal pattern of behavior. Thus, we are all "schooled" within each of our institutions, learning to conform to the supposed superiority of the powers that be. The problem for Illich, however, is the gap between our ideals and the actuality within any given institution. Thus, for example, there are the gaps between learning and teaching, between education and grade advancement, between competence and a diploma, and between the ability to say something new and a diploma. The student thus learns to increase his or her gap between aspirations and their fulfillment and also learns to bury any awareness of that gap so that it becomes more and more difficult to confront it effectively and narrow it. Further—and this is where Illich's analysis has much to teach us—this experience within the educational institution prepares the student to do the same thing within every other institution throughout society. Thus, what Illich describes is a process not only for conforming to a stratified worldview that fosters an ever-increasing aspirations-fulfillment gap but also how a cultural worldview or metaphysical stance maintains itself intact by teaching everyone to adopt that worldview.

From Illich's perspective, the usual political divide between those states that are more and those that are less democratic comes to look much less like a big divide, for they all create what he calls "schooled societies." Granting that there are indeed important differences between a society with political freedom along with many other kinds of freedom versus a society with far less freedom, Illich points up the failures of democratic societies due to their stratified nature. His argument, however, fails to account for the overriding importance of cultural values such as equality. It is within the nation-states that are more democratic and have more freedom that those cultural values are emphasized to the greatest extent. Such nation-states are further away from a "schooled society" than others. In more democratic countries people like Illich are able to voice critical opinions that alert us to our large aspirations-fulfillment gap. It is there that we have the greatest opportunities for confronting those gaps effectively and thus narrowing them. Yet piecemeal efforts at solution will not go far if our worldview or metaphysical stance remains stratified. Illich alerts us to a fundamental problem throughout contemporary society. Such a broad problem appears to require not only piecemeal efforts but also efforts to understand and alter the fundamental assumptions that we all make that are the basis for "a schooled society."

Researchers within sociology fail to emphasize the phenomenon of social stratification throughout all of our institutions—as did Illich—focusing instead on the political and economic institutions, and thus moving away from an examination of the metaphysical stance or fundamental assumptions throughout contemporary society. Nevertheless, we have much to learn by examining that literature, for it can give us greater insight into the nature of our stratified worldview. For example, the generally adopted definition of "power" is the ability to control the behavior of others, against their will if necessary. This outward orientation is a reflection of the outward perception discussed in chapter 2—by contrast with inward-outward perception—that is characteristic of a stratified worldview. By focusing on power over others, we fail to become aware of the importance of the individual's development of his or her own inner resources, such as learning to feel more confident and learning to communicate more effectively. For that kind of education will in turn enable the individual to achieve greater power over others. Of course, this is psychological development, and the orientation of each of the social sciences emphasizes its own supposedly limited territory: sociologists, thus, must confine themselves largely to social structures and must not be concerned with personality structures. Of course, this prevents sociologists from going very far in learning how social structures actually work.

With respect to the political institution, sociological analyses emphasize three different theories as to the distribution of power. We have the "power-structure approach," based largely on the work of Karl Marx, which focuses on the overriding importance of the economic institution for an understanding of power in society. Within the economic institution it is the owners of the means of production, or the bourgeoisie, who are the social class wielding the power in society. The workers, or the proletariat—the other major social class—have relatively little power, so long as they learn to accept their situation rather than become aware of it and work together to change it. The "power-elite approach" is based on *The Power Elite* (1956), the widely cited book by C. Wright Mills. He distinguished among three levels of power in American society: (1) the power elite (who hold top positions in government, major corporations, and the military, and who include the most famous celebrities); (2) the middle levels of power (members of Congress, key governmental officials, corporate and banking leaders, and other celebrities); and (3) the masses (who are relatively powerless). Mills thus saw several important hierarchies in addition to the economic one, by contrast with Marx. The third theory, the "pluralist approach," was developed largely by David Riesman (1961). It focuses on the competition among elites, maintaining that no elite can attain complete dominance, and that elites check and balance one another, following the ideals of a democratic state. All three approaches recognize the importance of social stratification, but they differ in locating the power in society.

To illustrate a current analysis of the political institution in contemporary society, let us turn to an article by G. William Domhoff (2006), where he reviews the adequacy of Mills's *The Power Elite* in light of events over the past 50 years since its publication. The book was roundly criticized by both Marxists and liberals, given its departure from the power-structure approach and the pluralist approach. Domhoff finds that "Mills was mostly right about the top levels of the power structure, but mostly wrong about the other levels of American society." For example, he "underestimated the power of Congress and too quickly dismissed the political parties as indistinguishable on power issues." And he failed to anticipate "a successful corporate counterattack that would reverse the gains made by organized labor." From a cross-national perspective, it appears to Domhoff—in addition to the inclusion within the power elite of economic, political, and military actors—that the elite should be extended to include religious actors, illustrated by the Iranian revolution and the Christian Right. Also, Domhoff does not see the masses as powerless, since—at least potentially—they forge a powerful coalition that has gained power, just as they have succeeded in accomplishing in the case of the civil rights movement, the antiwar movement, the women's movement, the environmental and consumer movements, and the gay and lesbian movement. Overall, then, it appears that Domhoff's approach largely rejects Marx's analysis in favor of a combination of Mills's and Riesman's analyses, although his sense of fundamental problems in modern society is much closer to that of Mills than to that of Riesman.

A fundamental problem with all three approaches to understanding power in society—including current analyses within the sociological literature like that of Domhoff—is that they focus on the distribution of power rather than on its genesis or development. Further, these approaches implicitly assume that the amount of available power at any one time is fixed, implying that they need not focus on the production of power but only on its distribution. This fixed perspective meshes with the static orientation of a stratified or bureaucratic worldview by contrast with an evolutionary or scientific one. It also illustrates the power of the former worldview to point the behavior of social scientists away from the scientific ideal of examining their own assumptions and challenging them when they prove to be so narrow as to sharply restrict their investigations or depart from realism.

Let us recall the definition of power that is widely accepted: the ability to control the behavior of others, against their will if necessary. To illustrate political development, we might turn to an analysis of the successful efforts of Mohandas Gandhi to free India from its status as a colony of Great Britain (Bondurant, 1965; see also Phillips, 1979: 283-286; Phillips and Johnston, 2007: 223-227). For example, responding to the British salt tax, which created hardships for poor people for whom salt was a necessity, Gandhi together with other leaders of the Indian National Congress led a civil disobedience campaign. They started with a 200-mile march to the sea from Ahmedabad to Dandi, where they would prepare salt from seawater. On March 2, 1930, Gandhi wrote a letter to Lord Irwin, the British viceroy, reviewing the grievances of the Indian people and informing him of the specific plans for the march. Those plans included constructive work undertaken by the volunteers during their journey, such as avoiding intoxicants and committing themselves to overcoming discrimination against untouchables. On April 5 the group reached Dandi, and on April 6 they broke the salt laws by preparing salt from the sea.

When leaders of the salt campaign were arrested, civil disobedience broke out throughout India, with the closing of shops, the nonpayment of taxes, and the resignation of large numbers of Indian officials. Upon hearing of outbreaks of violence in Karachi and Calcutta, Gandhi quickly took steps to curb those outbreaks. On April 17 Gandhi informed Lord Irwin of an intended march on Dharsana to occupy its salt works unless the salt tax was repealed. Gandhi was arrested on May 5. During that march volunteers arrived together with first-aid units, and many of them were repeatedly struck down by police batons without even trying to deflect the blows raining down on them. Publication of these events reached the British public, who increased their pressure on the government to repeal the salt tax. Other campaigns of civil disobedience throughout India followed, such as boycotts of British products, and these put increasing economic pressure on the British government. Given the threats from Nazi Germany and Japan during World War II, the Indian National Congress party did not push for independence until after that war, when it was finally granted.

Gandhi's techniques of civil disobedience followed at least these principles:

1. Refraining from any form of verbal or overt violence toward members of the rival group
2. Openly admitting to the rival group one's plans and intentions
3. Refraining from any action that will have the effect of humiliating the rival group
4. Making visible sacrifices for one's cause
5. Maintaining a consistent and persistent set of positive activities which are explicit (though partial) realizations of the group's objectives
6. Attempting to initiate direct personal interaction with members of the rival group, oriented toward engaging in friendly verbal discussions with them concerning the fundamental issues involved in the social struggle
7. Adopting a consistent attitude of trust toward the rival group and taking overt actions which demonstrate that one is, in fact, willing to act upon this attitude
8. Attempting to achieve a high degree of empathy with respect to the motives, affects, expectations, and attitudes of members of the rival group (Janis and Katz, 1959: 86)

These techniques all point away from a stratified worldview and toward an evolutionary worldview. We can also see them as procedures leading to Gandhi's development of power. For example, he did not force the volunteers who marched with him from Ahmedabad to Dandi to engage in constructive activities, like helping people who lived along that path to repair their houses and dig wells. He did not force them to break the salt laws when they reached Dandi. Neither did he force people throughout India to take up patterns of civil disobedience, nor did he force volunteers to risk their lives in their march on the Dharsana saltworks. He also did not force Lord Irwin into a compromise on the salt laws or the British government to grant India independence. Nevertheless, employing techniques such as in those listed, he succeeded in influencing people to follow his goals. Gandhi had no legal standing with the Indian or the British people, and he had no army at his command, but he developed the power to influence millions of people, even—for some—to the point of risking their lives.

Those principles are, of course, only part of the story of Gandhi's successes, yet they can yield insight when combined with the specifics of his activities. Overall, they reveal Gandhi's commitments to the people-oriented cultural values discussed in chapter 7—equality, freedom, individual personality, and democracy—values shared throughout India and Great Britain. Those commitments have to do with "head" and "heart." For example, with respect to "head," verbal abuse of the British occupiers was forbidden. As for "heart," Gandhi attempted to achieve an attitude of trust and empathy with them. In addition, Gandhi's hand or actions in support of his values and beliefs were crucial. Gandhi avoided overt violence and any action to humiliate the British. Also portraying his beliefs and ideals were his egalitarian and open relationship with Lord Irwin, his willingness to go to jail, his long fasts that brought him close to death, his willingness to lead the march to Dandi with the possibility of being clubbed to death, and much more. His assassination in 1948 by a Hindu fanatic came despite all of his efforts to achieve Hindu-Moslem unity. Unfortunately, the sociological literature on power generally ignores the process of developing power, illustrated by the life of Gandhi. That literature tends to ignore worldwide developments in favor of what has gone on in the United States and to ignore historical occurrences, such as the achievement of Indian independence. Thus, it tends to follow the orientation of a stratified or bureaucratic worldview to remain in Flatland rather than Spaceland or Timeland.

Let us look beyond the political institution to the economic institution to examine the nature of patterns of stratification throughout the modern world along with our stratified worldview. Economists generally define the economic institution as centering on the production, distribution, and use of wealth. Yet their approach to these processes is quite narrow, following a stratified worldview or metaphysical stance. A devastating criticism of that narrowness was mounted by E. F. Schumacher in *Small Is Beautiful* (1973), a book we encountered in chapter 6 that is far more relevant now when we attempt to face environmental crises that Schumacher warned us about:

> One of the most fateful errors of our age is the belief that "the problem of production" has been solved. Not only is this belief firmly held by people remote from production and therefore professionally unacquainted with the facts—it is held by virtually all the experts, the captains of industry, the economic managers in the governments of the world, the academic and not-so-academic economists, not to mention the economic journalists.... For the rich countries, they say, the most important task is "education for leisure" and, for the poor countries, the "transfer of technology."...
>
> A businessman would not consider a firm to have solved its problems of production and to have achieved viability if he saw that it was rapidly consuming its capital. How, then, could we overlook this vital fact when it comes to that very big firm, the economy of Spaceship Earth and, in particular, the economies of its rich passengers?
>
> One reason for overlooking this vital fact is that we are estranged from reality and inclined to treat as valueless everything that we have not made ourselves. Even the great Dr. Marx fell into this devastating error when he formulated the so-called "labour theory of value." Now, we have indeed laboured to make some of the capital which today helps us to produce—a large fund of scientific, technological, and other knowledge; an elaborate physical infrastructure; innumerable types of sophisticated capital equipment, etc.—but all this is but a small part of the total capital we are using. Far larger is the capital provided by nature and not by man—and we do not even recognise it as such. This larger part is now being used up at an alarming rate, and that is why it is an absurd and suicidal error to believe, and act on the belief, that the problem of production has been solved. (Schumacher, 1973: 12-14)

It was difficult for Schumacher to understand how people in business could overlook the vital fact that the economy of Spaceship Earth is rapidly consuming its capital, given that individual businessmen are very much aware of the dangers involved if they discovered that they were consuming the capital within their own businesses. Yet this is further evidence for the power of a stratified or bureaucratic worldview to shape our beliefs, our values, and our actions. As we saw in chapter 1, we are all largely residents of Flatland rather than Spaceland or Timeland, given the nature of our worldview. And as a result we have learned to see our situation from a narrow spatial and temporal perspective. This is the case even for those whom we see as having a broader perspective than almost everyone else, such as Karl Marx. He certainly had a very long-term approach to history, and he certainly looked ahead to a distant future. And he certainly opened up to a range of societies other than his own. Nevertheless, he along with everyone else appears to have been captured by a stratified or bureaucratic worldview. We can see this in the narrowness of his labor theory of value, as Schumacher argues. We can also see it in his failure to take culture very seriously, in his focus on the economic institution while largely excluding the importance of other institutions, and in his failure to understand the dangers of establishing a "dictatorship of the proletariat" as a "temporary" measure in moving toward a better society.

Beyond Schumacher's idea that the problem of production has not been solved, there is also the question of the nature of wealth, with economists almost invariably emphasizing material phenomena. Yet, as we have seen in chapter 6, there are cultural values other than material comfort. For example, there are people-oriented values like equality, freedom, democracy, and individual personality. Schumacher expresses the importance of such values, by contrast with the materialism stressed by economists, most forcefully:

> What is the meaning of democracy, freedom, human dignity, standard of living, self-realisation, fulfillment? Is it a matter of goods, or of people? Of course it is a matter of people.... If economic thinking ... cannot get beyond its vast abstractions, the national income, the rate of growth, capital/output ratio, input-output analysis, labour mobility, capital accumulation; if it cannot get beyond all this and make contact with the human realities of poverty, frustration, alienation, despair, breakdown, crime, escapism, stress, congestion, ugliness, and spiritual death, then let us scrap economics and start afresh. (Schumacher, 1973: 70)

The previous brief examination of the political and economic institutions of contemporary society suggests not only their failure to address basic social problems but also their central role in producing those problems. To illustrate further, a substantial body of literature on environmental problems and economic injustice has focused on "the treadmill of production" by worldwide economic institutions (see, for example, Schnaiberg, 1980; Schnaiberg and Gould, 1994; Gould, Schnaiberg, and Weinberg, 1996). Treadmill theory has clearly demonstrated the lack of sustainability of present-day global patterns of production (Gould, Pellow, and Schnaiberg, 2008). That theory has also demonstrated the global injustice—indicated by the increasing worldwide gap between the rich and the poor—linked to the global economy. Thus, there appears to be overwhelming evidence, following Schumacher, that "the problem of production" has not been solved. Despite such evidence, however, the treadmill of production is continuing on its merry way, with no social movement that is, as yet, powerful enough to alter its course.

This analysis also suggests the central role of a stratified or bureaucratic worldview that is shaping economic and political institutions so that they continually fail to solve the problems they are supposed to solve with respect to the production and distribution of wealth (defined broadly) and power (also defined broadly to include the development of power). Yet our continued commitment to that worldview is by no means inevitable. We have the capacity to move toward an alternative worldview that can teach us to address those social problems, given our status as human beings in possession of language and the scientific method. For example, we are not limited to emphasizing the production of material goods, linked to the successes of physical science along with its technologies. Social movements can become more effective in opposing the treadmill of production, making full use of available knowledge indicating its profound unsustainability and injustice. Breakthroughs in social science can yield increasingly effective social and personal technologies. And the result can be an increasingly balanced worldwide economy that moves toward achieving economic sustainability and justice. Thus, it seems likely that the problem of production can in fact be solved under certain conditions.

Conscious Evolution

"Conscious Evolution" suggests the distinction between biological evolution and this other kind of evolution. As we have seen in chapter 2, a change in our worldview pointing us toward perceiving ourselves as well as external phenomena may alter our patterns of perception, which we think of as biological phenomena. But we humans are in the possession of language and an understanding of the scientific method, and as a result we can continue to learn without any limits. Thus, we can learn to alter our patterns of perception just as we can alter ourselves physically, for example, through diet and exercise. Of course, we can shape ourselves with the aid of modern medicine, which has proven to be a most powerful technology based on the scientific method. Let us not forget the importance of education, seen broadly to include the mass media along with learning that occurs within every one of our institutions. Given such technologies as the mass media and the computer, we humans are now in a position to communicate extremely widely and rapidly throughout the world. The slow process of biological evolution that has brought us to where we are has taken billions of years. Now we are in a position to accelerate our evolution with unbelievable speed. We have seen in the foregoing section some of the fundamental problems that we face in doing so. Yet we have also seen in the foregoing chapters that we humans have the capacity to overcome those problems.

The discipline of psychology has given us a great deal of understanding about the individual human being, but—along with the other social sciences—it has also worked to distort our understanding of ourselves. Just as sociology has never been able to give us a realistic vision of a world with less and less stratification, so has psychology failed to give us a realistic vision of individuals consciously evolving in their thinking, feelings, and actions throughout their lives, granting some partial exceptions. Perhaps the most destructive invention of some psychologists is the idea that we humans vary greatly in our intelligence or fundamental capacity to learn. In a volume examining the nature of the IQ test and its supposed link to intelligence or the individual's basic capacity to learn, the editors conclude:

> It should be no surprise that at a time when the social policy of dominant groups is to curtail mobility, to thwart change, to block redistribution of power and wealth, the IQ test is once again hailed as a valid and objective tool. But IQ as a synonym for cognition is as inaccurate as is the faith in the relation of school, merit and opportunity itself. And for those who seek to assure opportunity, they must first demythologize the IQ test. (Gartner, Greer, and Riessman, 1974: 6)

The 77-page article in Gartner and others by Samuel Bowles and Herbert Gintis, which analyzes the research literature on the IQ test, concludes: "We have here attempted to speed up the process of demystification by showing that the purportedly 'scientific' empirical basis of credentialism and IQ-ism is false" (Bowles and Gintis, 1974: 84). Despite such analyses of the inadequacy of the IQ or any other test yet developed to measure basic intelligence or one's capacity to learn based on one's heredity, despite the failure of almost all researchers to claim that the IQ test is indeed a measure of basic intelligence, and despite the fact that IQ scores can be changed substantially, the idea widely persists that the IQ score is a measure of basic intelligence. Apparently, that belief is fostered by our stratified worldview, which remains invisible yet which works to teach us that hierarchies not only exist but also persist, and the result is social stratification. Yet studies of human biology generally point in a different direction, discovering no fundamental differences among human beings that result in differences in their capacity to learn. We might recall here the analysis of Stephen Jay Gould, as examined in chapter 2:

> We are, in a more than metaphorical sense, permanent children.... Many central features of our anatomy link us with fetal and juvenile stages of primates: small face, vaulted cranium and large brain in relation to body size, unrotated big toe, foramen magnum under the skull for correct orientation of the head in upright posture, primary distribution of hair on head, armpits and pubic areas.... In other mammals, exploration, play, and flexibility of behavior are qualities of juveniles, only rarely of adults. We retain not only the anatomical stamp of childhood, but its mental flexibility as well.... Humans are learning animals. (Gould, 1981: 333–334)

Another fundamental myth about human intelligence that is both widespread and devastating with respect to beliefs that we humans can continue to achieve conscious evolution throughout our lives is the idea that our most creative years occur in youth, and that our loss of brain cells and memory loss as we age is conclusive evidence that we lose intelligence in our later years. Yet recent research suggests otherwise, as indicated in the article "Older Really Is Wiser, Research Finds: Though Stereotyped as Feeble, Aging Brains Are Actually Finely Tuned":

> The aging brain is subject to ... trouble with short-term memory and rapidly switching attention.... But ... [a]n emerging body of research

> shows that a surprising array of mental functions hold up well into old age, while others actually get better. Vocabulary improves, as do other verbal abilities such as facility with synonyms and antonyms. Older brains are packed with more so-called expert knowledge—information relevant to your occupation or hobby....
>
> While younger brains solve problems step-by-step, older brains call on cognitive templates, those generic outlines of a problem and a solution that worked before. It's the feeling you get when you see that a new situation or problem belongs to a class of situations or problems you have encountered before, with the result that you don't have to attack them methodically....
>
> The benefits that come to the mind with age extend beyond thinking. They also include a greater ability to put yourself in another person's mind, empathizing and understanding his thought processes—emotional wisdom.... Elderly volunteers given a list of behaviors that describe a made-up person ... were more likely to infer correctly that the person was dishonest, kind or intelligent. (Begley, 2007: 1A, 17A)

Given that these ideas about the limitations of the human being—a hierarchy of intelligence and loss of intelligence as we age—are no more than myths, what can be said about possibilities for the further development or the evolution of the individual based not on biological but rather on conscious evolution? If indeed the individual has the capacity, based on language, to continue to evolve intellectually, emotionally, and in the effectiveness of his or her actions, how can we actually proceed to accomplish this? Before focusing on this, it is essential to clear away any focus on society as a whole rather than on the individual. For that focus has, up to this point, distracted us from the possibility of the individual's conscious evolution. We should understand that a focus on society or the group is part and parcel of our outer-oriented stratified worldview. My point here is a matter of taking first things first. Not that cultural change is unimportant, for—historically—culture has been absolutely fundamental to the shaping of the individual. But given the escalating problems of contemporary society, it appears that we must learn to reverse things: the individual must learn to change culture, and it is indeed important for at least some individuals to do so, since such cultural change will in turn work to accelerate desperately needed changes in individuals throughout the world. For it is the individual who can move very quickly to develop awareness of escalating problems throughout society with the help of the scientific method, following Figure I-2. It is the individual, learning to use that method in everyday life, who can learn to harness that awareness to confront problems ever more effectively, demonstrating to others the potential of a broad approach to the scientific method. The result, again following Figure I-3, can be a shift from a stratified to a cultural evolutionary worldview that, in turn, will work to narrow the aspirations-fulfillment gap or anomie throughout the modern world.

The failings of past efforts to focus on society or culture as a whole are suggested by Alexander Pope's *Essay on Criticism*: "A little learning is a dangerous thing; Drink deep, or taste not the Pierian spring." What we have had is a series of sociologists who have learned a little about Darwin's theory of evolution and proceeded to develop theories of human evolution with little regard for the enormous complexity of human behavior along with the range of knowledge throughout social science. As a result, they have constructed theories of social as distinct from biological evolution that conformed to their stratified worldview, that have not withstood the test of further scholarship within the academic world, and that have given a bad name to the very idea of social or individual evolution. Perhaps the individual most responsible for this state of affairs was Herbert Spencer, the English philosopher and sociologist who understood natural selection as the survival of the fittest, a process preserving favored races in the struggle for life.

Spencer's analysis provided ammunition for conservative politicians and some business leaders to limit any efforts to help the poor or to give livable wages to workers. After all, the process of evolution had found them to be "unfit," and any aid offered to them would work only to halt the overall process of continuing human evolution that requires the continuing develop of the "fit" at the expense of the "unfit." Just as many psychologists—along with most of the rest of us—had been captured by our stratified worldview and had come to believe in a hierarchy of intelligence, so had some sociologists come to believe in a hierarchy of the fit over the unfit. Needless to say, Spencer's theory meshed hand in glove with Hitler's ideology, which carried that idea one step further: to the slaughter of the unfit, including, not only Jews, gypsies, and those with physical defects, but also a great many others who departed from his ideal of Aryan perfection. Yet we should not blame Spencer for Hitler, since he would have vigorously opposed what Hitler did, as illustrated by his belief that "no one can be perfectly free till all are free." Further, Spencer is to be commended for his attempt—with all its faults—to carry forward Darwin's idea of biological evolution into the realm of society.

Since Spencer's time, a number of sociologists and anthropologists have returned to the idea of social evolution—most recently calling their efforts theories of modernization—attempting to avoid Spencer's distortion of Darwin, a distortion that defined fitness in such an extremely narrow way. Analysis of these theories lies beyond the scope of this book, although Piotr Sztompka has given us an excellent and detailed treatment of ideas that go back to those of Comte and Spencer and forward to Lenski and beyond (1993). More recently, Lenski's book (2005), Sanderson's book (2006), and the recent articles reviewing Lenski's book (*Contemporary Sociology*, 2006) can help bring Sztompka's analysis up to date. One thing that emerges from his critique and my own is the importance of a complex view of evolution, one that takes into account personality structure and the concrete situation along with physical and biological structures. My examination of conscious evolution in this chapter, with its focus on personality structure, is not meant to slight the importance of those other factors. I believe that conscious evolution is important because it has been neglected by sociologists and anthropologists interested in human evolution. And I believe it has been neglected because of a bias—emphasized by our stratified or bureaucratic worldview—in favor of the importance of social structures over personality structures.

The "image of the future" for the individual that I am putting forward goes back to the analysis of Buddhism in chapters 2 and 4 and, more specifically, to those ideas of an East-West synthesis. My effort here builds on those ideas by linking them to the modern work within psychology on reinforcement theory. An East-West synthesis links a Western outward orientation (emphasizing such cultural values as economic progress and a democratic state) with an Eastern inward orientation (emphasizing the individual's accumulating karma throughout life and learning to escape from dukkha). As a result of this joining of an outward and an inward orientation, we have the inward-outward approach of an evolutionary—by contrast with a stratified—worldview. I have argued that this synthesis requires a strategy that calls for the individual in everyday life to lower expectations in the short run so as to narrow his or her aspirations-fulfillment gap, following an Eastern orientation. That strategy also calls on the individual to continually raise both aspirations and their fulfillment in the long run, given a narrowed gap that enables the individual to function ever more effectively. Yet why should the individual be able to accomplish this? Can we really expect individuals throughout society to learn how to continue to evolve without any limit?

To illustrate, let us suppose that, as I proceed to write these words, I become vaguely conscious of a fear that I am unable to do well. For example, my own specific knowledge of the sociological and anthropological literature on evolution does not come even close to that of Lenski or Sztompka. Who am I to be writing about "conscious evolution" and attempting to build on their insights? Further, I am certainly no scholar of the works of Buddha and his followers. How dare I attempt to develop an East-West orientation that combines his ideas with Western ideas? For that matter, dare I presume to have insight into "Western ideas," with all of their diversity? Let's suppose, now, that I learn to welcome this fear and, instead of repressing it, I follow Sigmund Freud's and Karen Horney's general advice and bring it up to the surface of my awareness. My orientation to the importance of "heart," no less than "head" and "hand," teaches me the importance of learning to express my emotions, and the reflexive approach of a broad scientific method teaches me the importance of looking deeply into my personal problems. Further, the pendulum metaphor for the scientific method gives me the optimism I require to achieve this since it promises that I can overcome this fear when I bring it to the surface. I can also look to my self-image and realize that I am the kind of person who has done this successfully many times in the past. And still further, I can look forward to my evolution to the extent that

I continue to learn how to bring my fears up to the surface and learn to act effectively in the face of those fears. For all my life I have been attempting, with limited success, to learn to achieve ever greater expression of my emotions, given my own background as an outer-oriented Westerner and male with an academic history. And as a result of these efforts, I believe that I was able in the previous paragraphs to deal with the extremely complex idea of social or conscious evolution without slighting the work of others or pretending that I know more than I actually do know.

What I'm suggesting in general through this illustration is that it is indeed possible for any individual to learn how to bring to the surface negative emotions instead of burying those emotions and to follow high aspirations—such as completing this chapter more quickly—as I might have attempted to do. My understanding of my emotional limitations illustrates a lower level of aspiration: I must not assume that I can do more than I am able to accomplish with those limitations. Here, I am following Buddha by abandoning my "thirst" for rapid external accomplishments and, instead, looking inward. Yet as a result of this internal examination, I succeeded in feeling easier about those fears. Apparently, my head and heart came closer together. Using psychological terms, I was able to achieve positive reinforcement, and it was I—rather than someone else—who was able to give myself that reinforcement. In other words, I was able to reinforce myself, by contrast to depending on others to do that job. Given that reinforcement and my subsequent ability to write those paragraphs as well as these, my self-confidence has increased and, I hope, has affected my self-image in the long run. As a result I believe that I am in a position to continue to develop my ability to think, to express myself emotionally, and to act ever more effectively. This is what I mean by "conscious evolution": learning to move along a path that has to do with intellectual and emotional development as well as increasingly effective actions. It is a road that I believe anyone and everyone can learn to take, a road that I believe we must learn to take at this time in history.

In addition to a focus on present situations, we can also learn to do the same for past situations, as illustrated by the analysis of dreams. One dream that I have had several times has to do with my experiences as a sociology instructor. I am proceeding to give a lecture before a group of students when one of them in the back row, who was paying no attention, stands up and walks out through the highly visible front door. Soon, another student walks out, to be followed by others, until I'm left with an empty classroom. Another version of this dream is that I show up for class, but no student appears. I frantically contact the college office only to learn that I'm in the right place at the right time. I continue to wait, but no one shows up for class. Although at the times I had those dreams I quickly repressed them, I can now see them as examples of fears I had that are by no means unusual. And I can now see that my earlier repression of them conformed to the dictates of a stratified, outward-oriented worldview. By reliving those dreams at this point in time, however, I'm able to see them in a much different light. I now understand more fully the importance of analyzing such dreams as a way to learn to express my emotions.

This analysis of my dreams along with the preceding illustration of my fears while writing about conscious evolution are examples of a focus on the momentary situation, the focus of the fourth and final part of this book. Just as in the case of personality structures (part 3) and social structures (part 4), it is most useful for part 5 to begin with a figure that presents key concepts within the extraordinary language of social science that have been found useful. Just as with Figures 4-1 and 6-1, Figure 9-1 will present new concepts together with those that have already appeared in previous figures. Up to this point my emphasis throughout this book has been on structures, including physical, biological, personality, and social ones. Yet we are limited in how far we can go toward understanding how structures change unless we take up situations as well. Although we have already analyzed situations to some extent, as illustrated in the previous paragraphs, we have yet to focus on them and introduce concepts that have been developed by social scientists to gain insight into situations. Of course, we cannot afford to abandon our understanding of structures as we proceed. For it is a combination of our knowledge of structures and situations that will help us learn more fully about change within society and the individual, and this includes both momentary and long-term change.

Part V

The Situation

A focus on the situation is what we all emphasize in everyday life, for this is what our stratified or bureaucratic worldview teaches us to do. We tell stories of our experiences in situations to one another; we perceive our environmental situation from one moment to the next; we watch films portraying individuals in specific situations; and we read of situations in newspapers, magazines, and books. In all of these ways we have experiences that are similar somewhat to those of our pet dogs and cats, granted that they are sharply limited to what they can perceive directly that is going on at close range, and we are not so limited. It is that very difference that makes for the incredible complexity of human behavior. As we have seen in parts 2, and 3, and 4, we are products not only of physical and biological structures in common with our pets but also of complex personality and social structures, granted that our pets do develop in these directions to a limited extent. It is our complex language, not available to those pets, that is the basis for these structures and all of our complexity. In other words, we bring our personality and social structures—along with our physical and biological structures—into every situation that we experience. The chapters within part 5 use the extraordinary language of social science to invoke those structures as we proceed to analyze situations. This broad approach to structures and the situation—which follows the requirements of a scientific or evolutionary worldview—is a far cry from what goes in introductory textbooks within the various social sciences. In those books, generally, each new chapter is a thing unto itself, with very little building on earlier ideas.

Unfortunately, social scientists have not yet developed many concepts that are useful for probing the complexities of any given situation—such as emotional behavior—preferring instead to develop concepts bearing on structures such as those in Figure 6-1. Once again, we can see their behavior as a product of a stratified or bureaucratic worldview that works to narrow the social scientist's research emphasis. Thus, for example, we find sociologists emphasizing social structures; psychologists, personality structures; and anthropologists, culture, which may be seen as a type of social structure. Historians, by contrast, do focus on situations, but thus far they generally have failed to apply structural concepts to the situations that they examine. Yet when we apply structural concepts to situations, as suggested at the end of chapter 8, we move toward understanding how change occurs. For a new structure is based on a series of new situations, since structures are nothing more than phenomena that persist over time. Granting the limited number of important concepts that bear on the enormous complexity of the situation, we do have some, however, and those concepts can help us understand what is going on within a given scene. Figure 9-1—"Relationships among Basic Structures and Situational Elements within the Individual and Society"—presents the same structural concepts depicted in Figure 6-1. The new concepts bearing on the situation—derived from sociology and psychology—appear in the center of Figure 9-1, separated into columns under "heart," "head," and "hand."

Definitions of these new concepts appear in the glossary after chapter 11. A few words, however, might be helpful at this point. Most important is the idea that all the concepts have to do with behavior within a momentary situation by contrast with repetitive behavior. For example, if a momentary "social interaction" between two individuals is repeated over time, then we can invoke the concept of "group"—a concept referring to a social structure—to help us understand what is going on. But if the contact between two individuals does not recur, then "social interaction" applies. As for the two concepts bearing on "reinforcement," psychologists like to use the term "negative reinforcement" rather than "punishment," since the latter has many of its own consequences, whereas "negative reinforcement" often consists only of the withdrawal of "positive reinforcement." "Relative deprivation" is a type of negative reinforcement that has to do with an individual's feeling of unjustified loss or frustration relative to other people, who are seen as experiencing less of a loss or even a gain. A "stereotypical idea" refers to the assignment of fixed characteristics to a category of individuals or groups, such as African Americans, whites, females, males, the old, the young, the poor, the rich, Protestants, Jews, Catholics, Mexicans, the Irish. Such assignments sometimes become the basis for attitudes of prejudice and patterns of discrimination against individuals or groups with a given label, but this is not necessarily the case. Such attitudes and practices have to do with social and personality structures relating to "heart" and "hand," illustrating the importance of linking situational behavior with structures.

It is significant that we can divide the situational concepts within Figure 9-1 into those emphasizing "heart," "head," and "hand," just as we have done for the concepts having to do with personality and social struc-

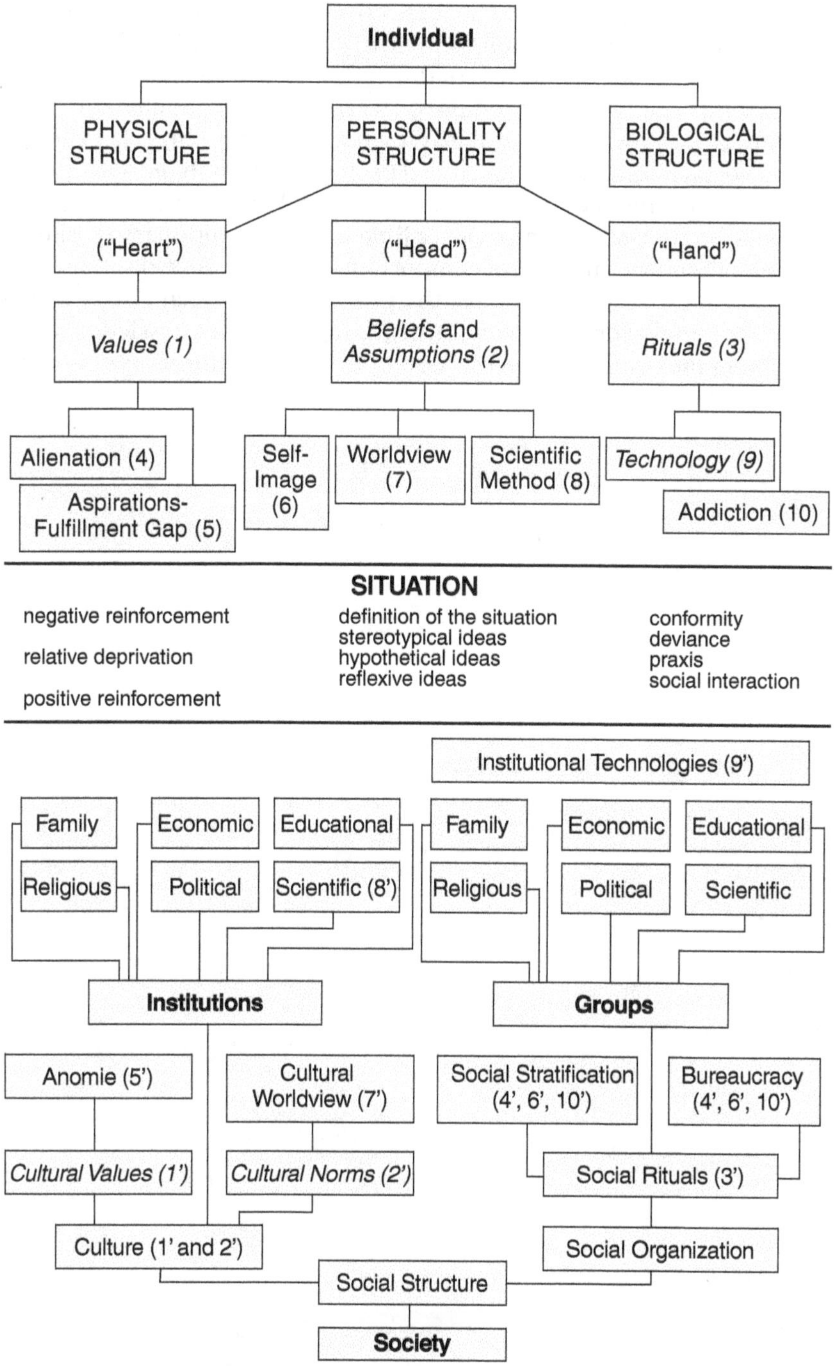

Figure 9-1. Relationships among Basic Structures and Situational Elements within the Individual and Society

tures structures in parts 3 and 4, respectively. Just as in those parts, one chapter will be devoted to each of the three elements of human behavior. I note once again the seriousness of the omission of concepts that can help us understand emotions—linked most clearly to "heart" in chapter 9—and their important role within any given situation. There are also different kinds of "social interaction"—discussed in chapter 11 on "hand"—with some concepts illustrating hierarchical behavior linked to a stratified worldview, and others tied to egalitarian behavior linked to an evolutionary worldview. Of course, there are many other important concepts developed by social scientists. Yet Figure 9-1 illustrates the potential that social scientists have for following the scientific ideal of integrating relevant knowledge around any problem they define, given the assumption of the enormous complexity of human behavior.

Chapter 9

"Heart"

Negative Reinforcement Versus Positive Reinforcement

Franklin D. Roosevelt—in his inaugural address on March 4, 1933—stated that "the only thing we have to fear is fear itself." By so doing, he brought to the surface of awareness the fear that a great many Americans during that era of the Great Depression were feeling. The president implied that the nation could move ahead without remaining paralyzed by fear and work to solve its severe economic problems. Although our stratified worldview calls on us to repress our fears, psychotherapists—including Freud and the neo-Freudians—generally suggest that it is far better to bring emotions like fear that we define negatively to the surface so that we can learn to work with them. What Roosevelt was doing in his speech was exactly that: making his audience more aware of the emotion of fear so that they would be able to confront it effectively. Unfortunately, however, it appears that our stratified worldview generally has the last word in teaching us to repress fear and other feelings that we have come to define as negative emotions. What yields a negative reinforcement is not the type of feelings we have but suppressing rather than becoming more aware of them so that we can function effectively and still accept those emotions. By pretending that we are fearless despite highly threatening situations, we tend to remain victims of fears that we have buried and thus refuse to recognize. In this chapter we begin with a section on negative reinforcement and end with one on positive reinforcement. Let us note that these are not sections on negative and positive emotions. All emotions can help us achieve positive reinforcement to the extent that we become aware of them and learn to work with them. And all emotions can yield negative reinforcement to the extent that we learn to keep them buried far from our awareness.

Negative Reinforcement

We begin with a study of depression, not to be confused with a negative mood, for clinical depression is a continuing and extremely negative state of mind that can easily lead to suicide. Although psychiatrists have tended to emphasize the experience of loss of a loved one as a key to understanding what causes depression, a new study published in *Archives in General Psychiatry* suggests the importance of humiliation as well:

> The events that send people into major depression, the authors found, are not merely losses, but humiliating ones that drive at a person's self-esteem—most typically, being abandoned by a romantic partner.... "When your father dies, it doesn't directly address who you are. You can still love yourself after your dad dies," said Dr. Kenneth Kendler, a psychiatric geneticist at Virginia Commonwealth University and the study's lead author. Blows to a person's status, like the experience of marital abandonment, "hit you at a more basic level.... So much of human life is built around this."
>
> Kendler followed 7,322 adult twins.... While the experience of loss put a twin at a higher risk for depression—about 10 times the risk of one who had not experienced loss—being humiliated was about as important. The two experiences together were by far the most dangerous kind, increasing the risk of major depression by a factor of 20. In one out of five cases, these individuals showed signs of major depression within a month....
>
> Kendler said his study's main contribution is the idea that humiliation, not pure loss, is a triggering event. The need to maintain one's self-respect may be more central to the personality than widely believed. "We are built to be status-protecting organisms," he said. (Barry, 2003)

Humiliation is an extreme form of shame. The study indicates the usefulness of tying situational occurrences to personality structures and, in particular, the individual's self-image, a concept to be found among the personality structures depicted in Figure 9-1. In this way, we are tying together a negative reinforcement—a feeling of humiliation—to the individual's persisting view, beliefs, or assumptions about self. Following Kendler's analysis, an individual experiencing both humiliation and loss is 20 times more likely to develop persisting depression than an individual who experiences neither. Given the close connection between humiliation and self-image, we can understand more easily the onset of depression. For self-image persists over time just as does depression, and a negative change in one's self-image appears likely to facilitate a long-term change in one's emotional state, such as a depression. Kendler's statement, "The need to maintain one's self-respect may be more central to the personality than widely believed," suggests the importance of linking self-image to the other concepts in Figure 9-1, just as was indicated in the introduction to part 5.

Some links were implied by my analysis of Scheff's study of shame, reported in chapter 3. Scheff, together with his colleague and wife, Suzanne Retzinger (1991), have come to see hidden versus acknowledged shame as fundamental to a range of modern problems. And they see hidden shame as increasing along with the industrialization and modernization of the past centuries. Yet why has this happened? If we refer to an increasing aspirations-fulfillment gap as we continue to modernize, then we can begin to understand the increasing repression of shame. If we had a direction for closing that gap, such as that offered by a broad approach to the scientific method, then there would be no need to repress our shame. Without such a direction, however, we have little choice. Instead of acknowledging our failures to ourselves and others, we bury our shame much like the Springdalers [reference to Vidich and Bensman's study, 1960, chapter 1]. And if that gap is increasing, then we will be constrained to increase our efforts to repress our shame at being unable to narrow that gap.

Assuming, then, that we are hiding shame more and more as modernization proceeds, I reasoned that this could be explained by our increasing aspirations-fulfillment gap, as depicted in Figure I-1. Since our stratified worldview gives us no direction for narrowing that gap but works to increase it, and since our scientific method is tied to that stratified worldview and offers little or no understanding of what is happening to us, then we have little choice but to bury our shame more and more as modernization proceeds with increasing that gap. This analysis carries further the link between shame, negative reinforcement, and self-image discussed earlier. It brings in the aspirations-fulfillment gap in society as a whole—or anomie—as well as the individual's aspirations-fulfillment gap. It also introduces the individual's worldview along with our cultural worldview. In addition, we have the scientific method put forward by the scientific institution. But we can find additional links to self-image as well. For example, the alienation and addiction of the individual are linked to the individual's stratified worldview. Patterns of social stratification and bureaucracy are tied to our stratified cultural worldview, which in turn shapes our individual worldview and affects our self-image negatively.

Scheff sees "pride" as, potentially, quite the opposite of shame, suggesting a self-image where the individual feels very good about self. But there is a serious problem with using "pride" in this way, a problem most likely brought on by our stratified worldview:

> Another issue concerns the meaning assigned to pride in the Old Testament. Virtually every reference places pride in a disparaging light. Perhaps the most familiar example occurs in Proverbs 16:18: "Pride goeth before destruction, and a haughty spirit before a fall." (In everyday usage, this

> quotation is often shortened to "Pride goeth before a fall.") A similar use occurs in 16:19: "Better it is to be of a humble spirit with the lowly than to divide the spoil with the proud" ... the word *pride* alone often carries a negative connotation. To escape that connotation, it is necessary to add a qualifier, e.g., *justified* pride.... In this usage, it would be necessary to add a qualifier, e.g., *false* pride, to refer to the kind of pride so prominent in the Old Testament.... False pride corresponds exactly to the meaning of the Greek word *hubris,* the kind of pride that leads certainly to nemesis, i.e., punishment by the gods.... In contrast to hubris, Aristotle depicted another kind of pride, which he saw as the supreme virtue, interest in honor above all else, a pride that means "greatness of soul." (Scheff, 1994: 43–44)

If we followed Aristotle's meaning for pride rather than the Greek meaning of *hubris* or the meaning of the Old Testament's haughty spirit, then pride could be seen as a positive, even evolutionary direction for our self-image. Yet ordinary usage stays close to that of the Old Testament and the Greek idea of *hubris.* Scheff's concept of justified pride has not come into general usage.

Scheff's approach to the scientific method, which he called the "Part/Whole" approach, was so close to my own that we later joined forces, and my former "Web" approach became the Web and Part/Whole Approach. Just as Figures 6-1 and 9-1 link personality structures to social structures, so does Scheff do the same by linking Hitler's personality to German social structure and change following World War I to help explain the origins of World War II. Germany's defeat in World War I, the humiliation of the Treaty of Versailles, the Great Inflation, and the Great Depression in 1929 created an economic, intellectual, political, and emotional crisis. Hitler himself, as he was growing up, appears to have been constantly in a state of anger bounded by shame, as illustrated by his continual temper tantrums. He would not listen to anyone and could fly into a rage over any triviality.

Scheff's analysis includes not only structures but also situational phenomena, as illustrated by his references to emotions. In these excerpts from Hitler's *Mein Kampf,* Scheff links humiliation or shame, pride, and hate or fury, showing us the power of Hitler's emotions and helping us understand his appeal to Germans who apparently were experiencing those emotions as well:

> There is ground for pride in our people only if we no longer need be ashamed of any class. But a people, half of which is wretched and careworn, or even deprived, offers so sorry a picture that no one should feel any pride in it. Only when a nation is healthy in all its members, in body and soul, can every man's joy in belonging to it rightly be magnified to that high sentiment which we designate as national pride. And this highest pride will only be felt by the man who knows the greatness of his nation....

> Particularly our German people which today lies broken and defenseless, exposed to the kicks of all the world, needs that suggestive force that lies in self-confidence. This self-confidence must be inculcated in the young national comrade from childhood on. His whole education and training must be so ordered as to give him the conviction that he is absolutely superior to others. Through his physical strength and dexterity, he must recover his faith in invincibility of his whole people. For what formerly led the German army to victory was the sum of the confidence which each individual had in himself and all together in their leadership. What will raise the German people up again is confidence....
>
> If at the beginning of the War and during the War twelve or fifteen thousand of these Hebrew corrupters of the people had been held under poison gas, as happened to hundreds of thousands of our very best German workers in the field, the sacrifice of millions at the front would not have been in vain. On the contrary, twelve thousand scoundrels eliminated in time might have saved the lives of a million real Germans, valuable for the future. (Scheff, 1994: 114, 116–117)

It is by using the concepts within Figure 9-1 that we can understand more clearly the power of Hitler's argument and his appeal to the German people. For one thing, there are his continuing references, direct and indirect, to the emotions of shame and pride, with both emotions running very deep and resulting in the fury we may note in the last paragraph. On the basis of a number of studies that Scheff cites, it appears that these same emotions were widespread among the German people, given the crises occurring in that country following World War I. Thus, Hitler's references to them must have sent out positive reinforcements. These emotions bear directly on the concept of self-image, referring to a positive self-image ("self-confidence" and "pride") and a negative self-image ("ashamed" and "broken"). As we have seen in Scheff and Retzinger's analysis of the hiding of shame in modern society, self-image is linked to a number of other concepts in Figure 9-1. For example, Hitler adopts a highly stratified approach when he writes about giving a youth "the conviction that he is absolutely superior to others." And his murderous tirade against Jews in the last paragraph exhibits extreme stratification linked to a stratified worldview or metaphysical stance. We can also detect a huge aspirations-fulfillment gap—much the same as a belief in the existence of anomie throughout Germany—running throughout these paragraphs. For example, the first paragraph presents an enormous contrast between "a people, half of which is wretched and careworn, or even deprived" and "a nation [that] is healthy in all its members, in body and soul," where there is "joy in belonging to it," and "national pride." Further, in the second paragraph Hitler contrasts "our German people which today lies broken and defenseless, exposed to the kicks of all the world" with the idea of educating the youth so that "he must recover his faith in invincibility of his whole people."

Our focus on negative reinforcements in relation to self-image can be extended to the phenomenon of "relative deprivation," which is a type of negative reinforcement involving an individual's feelings of unjustified loss relative to others whom the individual sees as experiencing less of a loss or even a gain. I referred to this concept in *Beyond Sociology's Tower of Babel* (2001: 105–113) as a "missing link" between personality structures and social structures. To illustrate let us return to our discussion of Levin's experiment on prejudice in chapter 4:

> The appearance of relative evaluation among Negro Americans may be having profound effects on the character of the civil rights movement in the United States and on the expectations which Negro Americans have about it. Even though the actual gains of Negroes have been occurring faster since 1940 than during any other period of Negro American history, the position of Negro Americans assumes a desperately deprived character when contrasted with the typical conditions of contemporary white Americans. . . . The result has been a keen sense of relative deprivation. (Levin, 1968: 26)

To the extent that African Americans compare their gains toward equality—whether in the educational, workplace, or any other situation—with their own past situations, they will experience a positive reinforcement that will tend to affect their self-images positively. To the extent that they compare their situation with those of whites, however, the result will tend to be a negative reinforcement with a negative impact on their self-images. Levin argued that it is the latter experience rather than the former that generally occurs, for people are "relative evaluators" rather than "self-evaluators," comparing themselves with others rather than with their own previous performances. My argument in chapter 4 did not focus on the situation but rather on personality structures. I concluded that people almost universally emphasize relative evaluation over self-evaluation, indicating that our worldview is stratified or bureaucratic rather than evolutionary or scientific. Given our continuing frustrations as a result of a large aspirations-fulfillment gap, we are prone to attitudes of prejudice against minority groups. In other words, we behave as if we are on a seesaw in relation to many others rather than on a stairway that all of us can climb. Moving now to a situational perspective, we can understand how a feeling of relative deprivation within a given situation can function as a missing link between personality and social structures by affecting both the individual's self- image and his or her relationships with other individuals or groups. For that feeling generally will be regarded by the individual as a negative reinforcement with reference to self-image, tying in with patterns of social stratification along with a stratified worldview throughout society.

With this view of relative deprivation, we can see more easily the links tying situational concepts to structural concepts in Figure 9-1. Let me stay with Levin's experiment on prejudice and take the risk of repeating myself because that risk may be worth it in view of the conclusions we can reach. Thus, an individual whose self-image and overall personality structure are oriented to relative evaluation rather than self-evaluation also exemplifies someone with a stratified worldview. We can assume that he or she did not learn that worldview in isolation but as a result of living within a society and a culture where such a worldview prevailed. More generally, based on arguments throughout the preceding chapters, we might assume that this situation is a microcosm of contemporary societies in general. I might add that I am writing this at a time when Don Imus's reference to the women on Rutgers University's basketball team as "nappy-headed hos" is being circulated by the mass media, and the problems of racism and sexism—beyond those remarks—are being debated at length. For me, the crucial question—beyond whether Imus should be fired—is what can be done about racism, sexism, classism, ageism, ethnocentrism, "weightism," "beautyism," and many other isms that appear to be pervasive in modern society. The media appear to be focusing not on this general problem—most likely because media people can think of no direction for solution given their lack of understanding of the contributions of social science—but rather on the specific situational question of whether Imus should be fired. Thus, the structural problem of continuing prejudice throughout modern society is largely ignored.

Positive Reinforcement

Yet the Levin experiment does suggest such an alternative, as indicated in chapter 4. Although the students engaged in relative evaluation by comparing their performance with that of others, some of them compared themselves at least to some extent to their own previous performances. Levin called this group "self evaluators" even though they engaged in such behavior to a limited extent. Yet that proved sufficient to prevent them from increasing their level of prejudice when they were frustrated. Those self-evaluators appear, then, to suggest an alternative worldview, pointing away from stratification and toward evolution. They had nothing to gain by reacting negatively to a minority group, since their orientation was—at least to some extent—to learn how to improve their own performances. This "stairway" versus "seesaw" behavior appears to have been linked closely to their personality structures. Thus, we might expect it to continue in any number of future situations involving frustration. Of course, the students' degree of self-evaluation was limited. If they had emphasized self-evaluation to a greater extent, it is reasonable to believe that their degree of prejudice would have been reduced considerably.

Granting the overriding importance of personality structures in moving away from prejudice and toward an evolutionary worldview, the question remains as to how we all can learn to deal with situations and work toward developing of personality structures linked to an evolutionary worldview. At the beginning of this chapter, I referred to Roosevelt's efforts to help people overcome their fears during the Great Depression by bringing them up to the surface of awareness. Another example, this one taken from a sequence of two novels by A. E. Van Vogt—*The World of Null-A* (1945/1970) and *The Players of Null-A* (1948)—can clarify this important process of learning to bring emotions such as fear, shame, guilt, and hate to the surface. I wrote about those novels in *The Invisible Crisis of Contemporary Society* (Phillips and Johnston, 2007: 202-207), but here we are in a position to trace the impact of this procedure on personality and social structures, as depicted in Figure 9-1. Van Vogt was deeply influenced by Alfred Korzybski's *Science and Sanity* (1933), a book that focused on the potential of a broad approach to language to enable us to solve fundamental world problems like prejudice, aggression, and war. In an unusual approach to a novel, Van Vogt quoted Korzybski at the beginning of his chapters. For example:

> For the sake of sanity, use ET CETERA: When you say, "Mary is a good girl!" be aware that Mary is much more than "good." Mary is "good," nice, kind, et cetera, meaning she also has other characteristics. It is worth remembering, also, that modern psychology—1956—does not consider the placidly "good" individual a healthy personality. (Van Vogt, 1948: 89)

The quote refers to Korzybski's conviction—which I share—as to the incredible complexity of the individual. That complexity suggests a fundamental question: How can we proceed to understand human behavior, given that complexity? I argue that it becomes essential to build bridges linking the knowledge of human behavior contained within our many specialized fields. I further argue that illustrations like Figure 9-1, which depict relationships among the extraordinary concepts of social science, give us a direction for building those bridges. "Et Cetera" is a contemporary journal that attempts to carry forward Korzybski's ideas. His comment that a placidly "good" individual is not a healthy personality suggests that such an individual engages in a great deal of conformity to the dictates of others rather than developing a personality structure and self-image that makes choices based on his or her unique experiences. We shall take up this situational phenomenon of conformity in chapter 11, where we look to the "hand" or actions of the individual in a given situation. This phenomenon is particularly important because habitual conformity points away from an evolutionary personality structure and worldview toward a stratified personality structure and worldview.

Van Vogt's novels center on the quest of the hero, Gilbert Gosseyn—pronounced "go-sane" to suggest Van Vogt's and Korzybski's conviction that we live in an insane world and must learn to move toward sanity—to discover who he is and also to defeat Enro the Red, the villain who is attempting to conquer the galaxy. Gosseyn's struggle to discover his identity is a metaphor for the rest of us, just as is his name. Who are we human beings? Exactly what capacities do we possess? Gosseyn discovers that he has, potentially, tremendous powers, indicated by his extra brain, yet another metaphor, this time for the potential of the mental powers that we all have. But he must work very hard to learn how to actually develop and use his mental powers, which, of course, all of us must learn to do. What is our human potential? Just how far are we able to develop it? And how should we proceed to do so? Gosseyn's education is along the lines of an evolutionary worldview. The focus is ongoing beyond the two-valued or dichotomous orientation to language emphasized by Aristotle, rejecting its narrow and static nature, and moving toward the gradational approach to language emphasized by the scientific method. Hence the concept "Null-A" in the titles of both books: "non-Aristotelian." Korzybski's work emphasized the overriding importance of language for human beings, and Van Vogt moves in the same direction. My orientation is not to emphasize gradation over dichotomy but rather to see both as important and also to add a third dimension of language: metaphorical or figurative language. The latter not only captures the fundamental contribution of the humanities or arts to language but also enables us to focus on the momentary scene to a greater extent. In his treatment of Enro the Red, Van Vogt illustrates a stratified worldview, focusing on achieving a highly stratified and bureaucratic system throughout the galaxy, with himself as its dictator.

Returning to the example in these novels of a procedure that can help an individual raise emotions like fear and shame to the surface and thus work with those emotions constructively, we turn to Prince Ashargin. His home planet, Gorgzid, was taken over by Enro while Ashargin—heir to the throne of Gorgzid—was still a child. After Enro murdered Ashargin's parents and all of his relatives, he kept Ashargin alive in the event that he might later prove to be useful. Yet Ashargin had been brought up in an extremely menial position, much like a slave, having learned to conform to the dictates of his master and having become a creature with little ability to understand or express powerful emotions like fear, once again a metaphor for the rest of us living in our stratified and bureaucratic world with its anomie or huge aspirations-fulfillment gap. The story then moves to Enro's increasingly successful war with the Galactic League, where his huge fleet of warships is succeeding in taking over planet after planet and defeating the warships of the league.

At this fateful time in the history of the galaxy, Gosseyn's mind has somehow been transported into the body of Ashargin. For Gosseyn to act quickly and decisively to save the galaxy—trapped as he is in Ashargin's body—he must learn to transform a weak and conforming Ashargin into a figure powerful enough to oppose the commands of Enro, the supreme ruler of the galaxy, and help the Galactic League achieve victory over Enro's fleet. Further, he must do so rapidly, since the Galactic League—previously coached by Gosseyn—is about to launch a mammoth mental attack on the planet Gorgzid. But how should Gosseyn proceed to train Ashargin, and to train him at lightning speed? Here, Van Vogt is raising a question for the rest of us: How can we learn to develop our incredible potential as human beings?

Gosseyn—or rather Gosseyn-Ashargin—sees his situation as follows:

> The mind of Ashargin ... had been an unintegrated, insecure mind, dominated by fears and uncontrollable emotions that were imprinted on the nervous system and muscles of the body. The deadly part of that domination was that the living flesh of Ashargin would react to all that internal imbalance on the unconscious level. Even Gilbert Gosseyn, knowing what was wrong, would have scarcely any influence over those violent physical compulsions—until he could train the body of Ashargin to the *cortical-thalamic* sanity of Null-A. (Van Vogt, 1948: 24)

Gosseyn proceeds to give the prince a key idea: "Prince, every time you take a positive action on the basis of a high-level consideration, you establish certainties of courage, self-assurance and skills." Then Gosseyn follows this up with specific procedures:

> In the bedroom Gosseyn rigged up a wall recorder to repeat a three-minute relaxation pattern. Then he lay down. During the hour that followed he never quite went to sleep.... Lying there, he allowed his mind to idle around the harsher memories of Ashargin's prison years. Each time he came to an incident that had made a profound impression he talked silently to the younger Ashargin.... From his greater height of understanding, he assured the younger individual that the affective incident must be looked at from a different angle than that of a frightened youth. Assured him that fear of pain and fear of death were emotions that could be overcome, and that in short the shock incident which had once affected him so profoundly no longer had any meaning for him. More than that, in future he would have better understanding of such moments, and he would never again be affected in an adverse fashion. (Van Vogt, 1948: 115-116)

Crucial to such Null-A training is the "cortical-thalamic pause," which carries forward the idea of reflexive behavior. Such a pause is particularly important when an individual is under severe threat and is tempted to act solely on the basis of emotions like fear or shame rather than to make good use of intellectual capacities. The Galactic League suddenly launches its massive mental attack, and Gosseyn-Ashargin responds with his effort to achieve a cortical-thalamic pause:

> I am now relaxing ... and all stimuli are making the full circuit of my nervous system, along my spinal cord, to the thalamus, through the thalamus

> and up to the cortex, and through the cortex, and then, and only then, back through the thalamus and down into the nervous system. Always, I am consciously aware of the stimulus moving up to and through the cortex.
>
> That was the key. That was the difference between the Null-A superman and the animal man of the galaxy. The thalamus—the seat of emotions—and the cortex—the seat of discrimination—integrated, balanced in a warm and wonderful relationship. Emotions, not done away with, but made richer and more relaxed by the association with that part of the mind—the cortex—that could savor unnumbered subtle differences in the flow of feeling. All through the palace, men would be struggling in a developing panic against the powerful force that had struck at them. Once that panic began it would not stop short of hysteria.... Yet all that the individual had to do was to stop for an instant, and think: The stimulus is now going through my cortex. I'm thinking and feeling, not just feeling. (Van Vogt, 1948: 177–178)

The cortical-thalamic pause is, then, a procedure for reacting to a momentary situational threat that employs the full weight of one's intellect along with one's emotions, directly analogous to our own orientation to learning how to use the scientific method in everyday life. We might think of this pause as linking "head" and "heart" together, given that we generally keep them fairly separated—such as by our repression of powerful emotions—so that we remain unable to express our emotions. And this integration of "head" and "heart" enables the individual to employ his "hand" decisively, launching efforts to act to solve problems effectively. The pause, then, is a way that the individual, who has learned to separate "head," "heart," and "hand," can move on a path leading to his or her integrating them, and thus moving toward sanity. As the individual continues along this path, that integration becomes not merely an approach that must be evoked consciously in the cortical-thalamic pause. Rather, the individual achieves an integration of his personality and social structures so that "head," "heart," and "hand" come to be integrated without any special conscious effort. For the individual has altered those structures, and they act in an integrated way from one scene to another.

Yet it is one thing to write compelling fiction and quite another to write compelling nonfiction. We might think of Van Vogt's novels as emphasizing the "heart" while giving us vague hints about the "head." We require the achievements of social science, however—as illustrated by Figure 9-1—to carry those hints further. By so doing we can follow Korzybski's and Van Vogt's emphasis on the centrality of language for solving human problems. Of course, all I can do here is indicate how concepts in Figure 9-1 can be linked to Van Vogt's narrative. The real problem that we all face if indeed we wish to learn how to confront effectively the escalating problems of modern society is to convert our understanding here—which is much like a cortical-thalamic pause—into changes in our personality and social structures. Such changes require far more than an intellectual understanding at this point. They require learning to employ this broad approach to the scientific method in more and more scenes within our everyday lives. As a result, we will gradually shape our personality and social structures so that we will—in Korzybski's and Van Vogt's terms—move ever closer to sanity. We will learn to pull ourselves together and integrate "head," "heart," and "hand" so that special situational efforts to do so (like the cortical-thalamic pause) will be less and less necessary.

Looking to Figure 9-1, the cortical-thalamic pause in general works to integrate personality and social structures so that the individual can apply them quickly within a momentary situation, for speed is required by what occurs in any given scene. Here we are integrating Figure 9-1 vertically. But that pause also integrates Figure 9-1 horizontally, since it works to integrate the columns of "head," "heart," and "hand." In addition to this general analysis, we can look to specific concepts within Figure 9-1 that illustrate elements of the cortical-thalamic pause. We might start with a "self-image" tying "head," "heart," and "hand" together, achieved to an increasing extent by our situational behavior—such as avoiding "negative reinforcements" like feelings of "relative deprivation" and emphasizing "positive reinforcements"—that, over time, structures our "self-image." As we shall learn in chapters 10 and 11, such situational behavior includes our "definition of the situation," the way we "label" others and ourselves, and our practices of "conformity," "deviance," and "social interaction."

In addition, every other concept listed in Figure 9-1 comes into play within our cortical-thalamic pause. For what is being sketched here is nothing less than a shift from a stratified to an evolutionary "worldview," based on our learning to use a very broad approach to the "scientific method" in everyday life. As a result of our learning to apply this scientific method—which is a social "technology"—we learn to narrow our "aspirations-fulfillment gap" and reduce our "alienation" and "addiction." To achieve all of this, it is also essential that we alter more and more of our momentary perceptions in an inward-outward (versus outward) direction, thus taking into account a widening range of physical structures and moving from Flatland into Timeland and Spaceland. Up to this point, I have invoked concepts bearing on personality, physical and biological structures along with those bearing on the situation. Yet such structures affecting the individual's behavior would have an immediate impact on the small-scale social structures surrounding him or her, pointing away from anomie, a stratified worldview, and also patterns of social stratification and bureaucracy.

As just indicated, being able to achieve a cortical-thalamic pause within a given scene so as to be able to alter our personality and social structures is an enormous undertaking. Yet we can proceed to move in that direction by looking at additional illustrations that focus on other scenes where we can once again achieve a cortical-thalamic pause. One such scene is presented by Laura Sperling, a columnist, on the op-ed page of the *Sarasota Herald-Tribune*:

> Regular viewers of the terrorism drama "24" know better than to watch the show right before bedtime. But that's what I did Monday, foolishly thinking the over-adrenalized jolt of car chases, torture and evil political intrigue would balance out a calm two-hour Sarasota City Commission candidate forum I had sat through beforehand. Instead, the two experiences merged in my head and sparked a night of insomnia....
>
> Yet Chinese water torture—drip ... drip ... drip—did come to mind when I thought of the way local governments attack long-running problems. Other voters must feel the same, because at the forum they raised many questions about slow-moving civic issues. Since 2002, for example, Sarasota has had a mostly unfunded plan to revitalize low-income Newtown.... Similarly, it's been more than six years since roundabouts were recommended for the critically traffic-challenged stretch of U.S. 41 near the Ritz-Carlton hotel ... Drip ... drip ... drip ...
>
> As these and other examples drearily splashed through my mind the other night, they contrasted with the hyperdrive images of Jack Bauer—fictional governmental agent of "24"—blasting his way through Monday's TV torment. Squeezing the most from every available second, Jack—the quintessential man on a mission—finds a way to disarm a terrorist's nuclear suitcase bomb. He's assisted by a super-efficient crew of geeks who man the computers back at "CTU," a government agency with such an all-knowing and seamlessly linked array of spy satellites and databases that even the eyes of God would look on in envy.
>
> Locked in insomnia, I wondered: what if municipal governments were like this—so effective and mission-minded that they measured their achievements in seconds and hours rather than years and decades? ... Don't try this at home, folks; last time I interfaced with Sarasota's trash-collection office, it took several phone calls and a week's wait to accomplish my mission (replacing a missing garbage can).... But in exchange for the peace that reigns in Southwest Florida, must we accept the status quo, with all its faults? Without something to serve as motivation, what will spawn the courage, conviction and exhaustive effort it takes to solve longstanding community problems? Urgency is clear enough when a nuclear bomb stares us in the face. But what—besides drip, drip, drip—can we do the rest of the time? (Sperling, 2007: 19A)

Once again we can link a narrative of events within momentary situations to Figure 9-1 to achieve another cortical-thalamic pause. This time the narrative is not fictional, but it still lacks any use of the extraordinary concepts of social science. The fundamental question that Sperling poses is whether it is possible for us humans—faced with social problems and moving very slowly toward solving them—somehow to learn to move with the speed and effectiveness of Jack Bauer or, for that matter, Gilbert Gosseyn, who moves across the galaxy at a speed far exceeding that of light. Adding to Sperling's question, we might also wonder whether we can learn to act with such speed to confront effectively the far more threatening problems facing the human race at this time in history, such as biological, chemical, and nuclear terrorism; global warming; and our escalating aspirations-fulfillment gap.

Yet by posing this question in a way that communicates most powerfully, Sperling succeeds in taking a giant step toward making progress on all of these problems. She begins to take the first and most important step of the "scientific method": becoming aware of and being committed to solving a problem. She is suggesting that we develop an effective social "technology" for solving Sarasota's problems of housing, traffic, and garbage collection. While posing such problems, she is voicing her "aspirations-fulfillment gap"—between aspirations for solutions and the city's present failures—instead of burying those aspirations with the fatalistic saw that "you can't change city hall." I would argue further that it takes a certain kind of individual to accomplish this: someone with a "self-image" and "personality structure" who has managed to avoid "alienation" and "addiction" to a fair extent, who has managed to deal with her "biological structure" and "physical structure" by moving into Timeland and Spaceland to some extent, and who has also moved toward an evolutionary "worldview."

Sperling's comparison of Jack Bauer's speed of action with that of the Sarasota City Commission invokes attention to the speed of events within any momentary scene. What we humans appear to require are key aspects of the personality structure of a Jack Bauer or Gilbert Gosseyn, and not our present personality structures that divide "head," "heart," and "hand" in one scene after another. We need to learn how to carry further Sperling's movement in this direction. Apparently, she caught a glimpse of what would be involved in the behavior of Jack Bauer, just as I caught a glimpse of it in the behavior of Gilbert Gosseyn.

What is most encouraging about Sperling's ideas is her implication that we need not wait till the problems of the world have been solved to take action locally. I would argue, in addition, that such local action can pave the way for movement toward addressing world problems, following the pendulum metaphor for the scientific method. This requires a vision of a different kind of world, such as one where we can move toward developing elements of the personality structures of Jack Bauer or Gilbert Gosseyn. This is indeed the world that I have attempted to sketch in this book, with my own vision of an evolutionary or scientific world by contrast with our present stratified and bureaucratic one. With such a vision we can learn to use the scientific method—as indicated by the pendulum figure of speech or metaphor—toward a deeper sense of problem no less urgent than that felt by Jack Bauer and Gilbert Gosseyn. Given the momentum that having a sense of problem can yield, we can move further toward confronting our fundamental problems effectively.

While chapter 9 has treated situations emphasizing the "heart," chapter 10 presents scenes centering on the "head" by clarifying the concepts "definition of the situation," "stereotypical ideas," "hypothetical ideas." and "reflexive ideas," with no assumption that these concepts are sufficient by themselves to take us beyond a certain point in understanding this topic. We might note that all of these concepts emphasize language, which is absolutely fundamental to human behavior and human problems. For example, it is written language that has made possible the scientist's ability to "stand on the shoulders of giants," building on the studies of those who have gone before. I might argue that social science's greatest achievement in the twentieth century was the discovery of the overwhelming importance of language for understanding the nature of our human situation and our problems. Given that insight, we should be prepared for a next step: learning to use language—just as Korzybski and Van Vogt envisioned—so that we humans can learn to confront our escalating problems ever more effectively. In my view, this requires learning to use the extraordinary language of social science—as sketched in Figure 9-1—in everyday life. I might also note that "reflexive ideas" point us squarely toward an evolutionary worldview. It is a key idea that distinguishes the Web and Part/Whole Approach to the scientific method—the basis for my approach throughout this book—from the traditional approach.

CHAPTER 10

"Head"

Stereotypical Versus Hypothetical Ideas

IN INTRODUCING CHAPTER 10 WITH ITS EMPHASIS ON THE SITUATION and on language, let us begin by clarifying the four situational concepts located under the "head" column: "definition of the situation," "stereotypical ideas," "hypothetical ideas," and "reflexive ideas." The "definition of the situation" is a concept developed by a social psychologist, William I. Thomas, in his effort to study the behavior of young women who have problems in adjusting to their overall situation inside and outside their families (Thomas, 1923). We may understand this concept in a very general way as referring to any understanding of the nature of a momentary situation. For example, we might see a given situation as limited to whatever we can actually sense with our sight, hearing, and other senses, corresponding to the narrowness of a Flatland orientation, as discussed in chapter 1 and as linked to a stratified or bureaucratic worldview. By contrast, we might understand a given scene as including—in addition to what we can sense—the entire history of the universe, given that any scene is indirectly affected by every other scene within our interactive universe. This definition of the situation corresponds to the breadth of a Spaceland and Timeland orientation—as discussed in chapter 1—and is linked to an evolutionary or scientific worldview. Alternatively, we might define a situation as somewhere in between these two widely different definitions, such as including some or all of the past experiences of the individuals involved in the scene.

"Definition of the situation" thus alerts us to the idea that we have a choice to make within any and every situation in which we are involved. As just suggested, we can choose a definition that is very narrow, very broad, or intermediate with respect to the range of phenomena involved. We can thus choose to view a situation from the perspective of a stratified worldview, an evolutionary worldview, or from somewhere in between. My argument in the preceding chapters has been that we actually generally follow a stratified or bureaucratic worldview in our choices, thus yielding the escalating aspirations-fulfillment gap that we see in Figure I-1 and resulting in choices that define the situation narrowly. For example, in chapter 2 the black hero in Ralph Ellison's (1947) *Invisible Man* was defined stereotypically in his situational encounters with most whites. He was labeled as a black, and there was little interest in any other of his characteristics as a human being. Similarly, the rest of us tend to see one another and ourselves in narrow ways relative to the incredible complexity of the human being, as illustrated by the number of unique experiences we have all had in the past. We are all, then, invisible to a great degree. Yet I believe that we can learn to move toward an evolutionary or scientific worldview—accompanied by a broad scientific method that we use in everyday life, as sketched in Figure I-3—enabling us to define our momentary situations ever more broadly. We can thus learn to make one another and ourselves ever more visible. Conversely, to the extent that we can learn to define situations more broadly and make ourselves more visible, that will help us in turn move from a stratified to an evolutionary worldview.

"Stereotypical ideas" and "hypothetical ideas" refer to this same contrast between a narrow and stratified worldview versus a broad and evolutionary worldview, respectively. They help take us from the very general question of how we choose to define a situation to the specific linguistic tools we use to make that choice. George Kelly, one of the few psychologists who has focused on language, developed the metaphor of "man-the-scientist," as noted at the beginning of chapter 4. He attempted to develop a scientific approach to our everyday usage of language, just as did Alfred Korzybski and A. E. Van Vogt (chapter 9). Kelly examined this contrast between narrow and broad ideas, using the terms "pre-emptive constructs" and "constellatory constructs" to refer to stereotypical ideas and "propositional constructs" to refer to hypothetical ideas:

> Kelly ... classified constructs ... into *pre-emptive, constellatory* and *propositional*. A pre-emptive construct is a construct which pre-empts its elements for membership in its own realm exclusively. This is in effect saying that if this man is a homosexual he is *nothing but* a homosexual ... it is essentially a denial of the right of other people and ourselves to re-view, re-interpret and see in a fresh light some part of the world around us.
>
> A *constellatory* construct is a construct which fixes the other realm membership of its elements. This is essentially stereotyped or typological thinking and says in effect that if this man is a *homosexual* then he must be *effeminate, artistic, degenerate* and *a menace to society*. Again, it reduces our chances of elaborating or re-viewing our outlook—it is a kind of intellectual package deal.
>
> Finally, Kelly talked of *propositional* constructs, which are those constructs which carry no implications regarding the other realm membership

> of their elements. These are ... constructs where we are prepared to recognize that we can look upon person X *as a homosexual* and thereby make sense out of a lot of what he says and does. But we are recognizing that this is only *one* way of viewing him and is not some final, absolute or all-comprehending truth. We can equally regard him *as a friend* or *as a chess player*. The more propositional our constructs the richer becomes our world and the less likely we are to become irretrievably trapped into a conflict which arises out of the rigidity of our viewpoint. (Bannister and Fransella, 1971: 31–32; see also Phillips and Johnston, 2007: 47–52)

For Kelly, preemptive and constellatory constructs are unscientific in that they give us a narrow and fixed view of phenomena backed up by little or no evidence. When we apply them to individuals, such as stereotyping homosexuals as "artistic," we generally fail to take into account the range of available evidence on the subject. We also generally remain fixed in our belief and fail to be open to the full range of the personal characteristics of homosexuals so that a given homosexual individual comes to be seen as a unique human being. In other words, we see such individuals in a prejudicial way, where we "prejudge" who they are, leaving an individual's actual nature invisible, just as in the case of Ellison's invisible hero. A hypothetical construct, by contrast, is much like a scientific hypothesis. It is put forward tentatively, since there are no absolutely certain conclusions within the scientific method. It is, ideally, based on whatever evidence is presently available and is wide open to the development of new evidence that might change it or even go against it. Of course, we cannot stop to check for all the available scientific evidence that bears on every thought that we have or every choice that we make in everyday life. Yet to the extent that we move toward hypothetical versus stereotypical ideas, we point ourselves in a scientific direction, and we do the best that we can within any given situation. I argue that we can learn to do far better than what we do at present by moving from our stratified worldview to an evolutionary worldview and also by using a broad approach to the scientific method in everyday life, following Figure I-2.

The concept "reflexive ideas" in Figure 9-1 refers to inward-looking as distinct from outward-looking ideas within any given situation, such as an individual's thinking of how some past personal experience—such as changing lanes while driving without looking carefully—bears on the individual's present situation. It may appear that we are all quite reflexive, and many would argue that thinking too much about ourselves is the kind of "selfish" behavior that we should avoid. Others might argue that such thoughts generally would interfere with paying sufficient attention to the business at hand in any given scene. Yet my argument is not that reflexive ideas should substitute for attention to external phenomena so that we fail to pay attention to what we must do in the momentary situation. Rather, by remembering a past experience, for example, when we changed lanes without looking carefully, we can learn to look carefully at the present situation before changing lanes. Thus, I argue that reflexive ideas can help us bring our past experiences into the present moment so that we can take them into account. This is much like the behavior of the scientist who must first "review the literature," learning the results of previous studies before initiating new research.

Further, my argument—as developed in chapter 2's contrast between "invisible man" and "visible man"—is that we are largely invisible to ourselves. We have a long way to go in learning to substitute "hypothetical ideas"—that take into account the available evidence from our own past experiences—for "stereotypical ideas." Following Gouldner's ideas, as presented in the introduction and in chapter 1, sociologists fail to be reflexive so that they do not learn to apply their findings about human behavior to themselves. We can even go back as far as the New Testament, where we have the advice "Physician, heal thyself" (Luke 4: 23). Given the prevalence of patterns of stratification and bureaucracy throughout our institutions, I argue that we have learned to emphasize an outward orientation versus an inward-outward orientation. For example, Levin's experiment in chapter 4 contrasts "relative evaluation" with "self evaluation," where those individuals who combined the two orientations to at least some extent did not increase their prejudice against Puerto Ricans when they were frustrated. Yet those who emphasized the outward perspective of comparing themselves with others acted when frustrated as if they were on a seesaw in relation to Puerto Ricans, putting them down so that they could rise relative to them. Selfish behavior is focusing on oneself to the exclusion of others. My argument for the importance of reflexive ideas is that we redress our one-sided, outer-oriented worldview and take into account *both* ourselves and others. Instead of remaining on a seesaw with outward orientation that leads us to treat others in a stratified way, we might learn to see ourselves on a stairway taking us toward human development. It can be a stairway with extremely wide steps, so that there is also room for everyone else to climb it. And we can learn how to climb by looking at the experiences of our fellow climbers, just as they can learn from our own experiences.

"Reflexive ideas," noting the description of the Web and Part/Whole Approach in chapter 1, are central to a broad approach to the scientific method that follows an evolutionary versus a stratified worldview. By contrast, social scientists almost invariably have failed to assess their own impact on the research process, as we can see throughout their many publications. This is by no means a conscious and willful effort to depart from scientific ideals, but rather a product of an outward-oriented or stratified worldview that has shaped their understanding of the scientific method. That worldview emphasizes the simplicity of human behavior rather than its incredible complexity. As a result, it is easy for a social scientist to become convinced that he or she need not dip into the complexity that would accompany assessment of his or her impact on the research process.

In this chapter we begin with a section on stereotypical ideas, continuing to focus on the momentary situation. Necessarily, we must also focus on the nature of the language that we use to define any given situation. We must bear in mind the importance of linking situational phenomena with the structural phenomena sketched in Figure 9-1. Our second and final section is on hypothetical ideas. The question raised here is: How can we learn to move from stereotypical ideas to hypothetical ideas? Both sections build on the previous discussion in the introduction of the three potentials of language: dichotomous, gradational, and figurative or metaphorical, as emphasized within social science, biophysical science, and the humanities, respectively. All three are important in our efforts to answer this question, but as yet our educational institution has failed us in our understanding of them, let alone our ability to use them from moment to moment in everyday life.

Stereotypical Ideas

In this section and the final one, we shall turn to fiction—without excluding nonfiction—to help us gain insight into the momentary situation, just as we have done in earlier chapters. This is not merely a question of putting icing on the cake of our analysis; fiction is essential for baking the cake itself. All language works to help us move up and down language's ladder of abstraction and to understand and solve problems, as illustrated in chapter 1's analysis of the scientific method. But it is fiction more than biophysical and social science that has emphasized the complex details within any given momentary situation. It is fiction that has illustrated, for example, the range of our emotions. And it is fiction that centers on language's metaphorical or figurative potential, just as physical science emphasizes its gradational potential and social science (and everyday thought) its dichotomous potential. For example, the reference to the icing on the cake and the cake itself are metaphors for something added to our understanding that is not fundamental (the icing) and something absolutely fundamental to our understanding (the cake).

Here we reference George Orwell's *Nineteen Eighty-Four* (1949; see also Phillips and Johnston, 2007: 208–210). It is known mainly for its nightmarish vision of Big Brother—whom Orwell modeled after Joseph Stalin—a dictator who, with the aid of two-way telescreens and the Thought Police, had developed a society far more totalitarian than any that has ever existed. Yet Orwell gave us another vision in his book, not well known, of the official language used in Big Brother's Oceania: Newspeak. He even included an appendix, "The Principles of Newspeak." Winston Smith, Orwell's hero, works in the Ministry of Truth that follows the party's three slogans: "WAR IS PEACE, FREEDOM IS SLAVERY, and IGNORANCE IS STRENGTH." These slogans give us a preview as to the nature of Newspeak, which is designed to promote them.

Syme, Winston's colleague in the Ministry of Truth, is working on compiling the definitive edition of the Newspeak dictionary. He confides what he is doing to Winston:

> You think, I dare say, that our chief job is inventing new words. But not a bit of it! We're destroying words—scores of them, hundreds of them every day.... Do you know that Newspeak is the only language in the world whose vocabulary gets smaller every year? ... Don't you see that the whole aim of Newspeak is to narrow the range of thought? In the end we shall make thoughtcrime literally impossible, because there will be no words in which to express it.... Every year fewer and fewer words, and the range of consciousness always a little smaller.... By 2050—earlier, probably—all real knowledge of Oldspeak will have disappeared. The whole literature of the past will have been destroyed. Chaucer, Shakespeare, Milton, Byron—they'll exist only in Newspeak versions, not merely changed into something different, but actually changed into something contradictory of what they used to be. Even the literature of the Party will change. Even the slogans will change. How could you have a slogan like "freedom is slavery" when the concept of freedom has been abolished? The whole climate of thought will be different. In fact there will be no thought, as we understand it now. Orthodoxy means not thinking—not needing to think. Orthodoxy is unconsciousness." (1949: 45-47)

It is our language that sharply distinguishes us from all other organisms and enables us to become fully human. Using language, we can achieve a given "definition of the situation" that integrates our unique past experiences with the momentary scene, thus helping us solve problems. Using language, we move toward penetrating the complexity of the behavior of others and ourselves. Yet it is "stereotypical ideas"—such as "freedom is slavery"—that are fixed in stone and support Big Brother's rule and that betray the potentials of language. Newspeak points in the direction of eliminating language and thus converting human beings into robots or sheep who automatically conform to Big Brother's wishes. As for "hypothetical ideas" that are tentative and suggest the importance of searching for evidence, those become creatures of the past in Oceania, where absolute certainty as to the righteousness of Big Brother prevails. Indeed, Winston Smith and others in the Ministry of Truth are busily rewriting all of the history books and all other records of the past so that they conform to Big Brother's views of what happened before his reign.

It is all too easy to read Orwell as someone deeply critical only of Stalin's Russia or of other dictatorships like Hitler's Germany, yet all the while failing to engage in "reflexive ideas" and look at the situation in the United States or Europe. Neil Postman, a psychologist who has written *Amusing Ourselves to Death: Public Discourse in the Age of Show Business* (1985), is convinced that Orwell's fears are actually realistic, but not as the result of a dictator like Big Brother. Instead, he follows Aldous Huxley's argument in *Brave New World* (1939). Postman believes that "no Big Brother is required to deprive people of their autonomy, maturity and history." As Huxley and Postman see it, "People will come to love their oppression, to adore the technologies that undo their capacities to think":

> What Orwell feared were those who would ban books. What Huxley feared was that there would be no reason to ban a book, for there would be no one who wanted to read one. Orwell feared those who would deprive us of information. Huxley feared those who would give us so much that we would be reduced to passivity and egoism. Orwell feared that the truth would be concealed from us. Huxley feared the truth would be drowned in a sea of irrelevance. Orwell feared we would become a captive culture. Huxley feared we would become a trivial culture.... In short, Orwell feared that what we hate will ruin us. Huxley feared that what we love will ruin us. This book is about the possibility that Huxley, not Orwell, was right. (Postman, 1985: vii–viii)

From the perspective of Figure 9-1, Postman and Huxley are arguing as to the overriding power of culture—including cultural values and norms, anomie, the cultural worldview, and all of our institutions—to shape the way we all think, feel, and act; and there is little question but that they are correct. For example, we might consider the power of television and the mass media in general to dumb down public debates to quick sound-bites that ignore the complexity of issues. Yet isn't Postman stereotyping Orwell's book by emphasizing his portrayal of Big Brother's patterns of "social organization"—including the extreme "social stratification" and "bureaucracy" in Oceania—while ignoring Orwell's emphasis on the importance of the cultural phenomenon of Newspeak? Language is the basis for culture, and Orwell deals with the range of social structures in his novel: culture as well as social organization. Further, if we look carefully at Huxley's *Brave New World,* we find that although he emphasizes culture he certainly does not ignore social organization. He gives us a vision of a population functioning as couch potatoes, yet he also describes the Controllers, who use eugenics and drugs to shape that passivity. For Orwell or Huxley to project a vision of society that makes sense to a reader, it is essential that they include descriptions of culture as well as patterns of social organization.

Given this attention to both social organization and culture, it is possible to strengthen Postman's argument—emphasizing culture and claiming that we are now moving into a Huxleyan world—by adding to it aspects of social organization that both Huxley and Orwell suggested. Suppose we introduce—in addition to our normal emphasis on dichotomous thought—a gradational orientation. Certainly Big Brother, Stalin, and Hitler were monstrous rulers, and certainly they slaughtered many millions of people. But what about seeing their dictatorships on a continuum with lesser degrees of social stratification, such as what exists in the Western world at this time in history with our stratified or bureaucratic worldview? What about the middle-class or working-class employee who trembles at the thought of losing his or her job because of a failure to conform to the exact wishes of a boss who cares little or nothing for the welfare of the worker? What about the social scientist who conforms to research procedures that are far removed from the ideals of the scientific method, yet who buries those ideals in the interest climbing higher on the ladder of success? What about the vast majority of the population of the Western world who have given in to cynicism and despair in the face of increasing personal and social problems that they and their elected leaders remain unable to solve? What about everyone who proceeds to escape from the responsibility of confronting those problems through the technique of particularization and the falsification of memory, leaving decision making in the hands of politicians and "experts" who are ignorant of how to understand those problems, let alone confront them effectively?

Further, we can also strengthen Postman's argument emphasizing our movement into a culture that is destroying our humanity by taking a closer look at Newspeak. Given the centrality of language in the development of human behavior, Orwell's appendix, "The Principles of Newspeak," gives us a rare opportunity to gain insight into just how language fails us, for Newspeak is all about the elimination of language. The twentieth century has taught us the importance of language, but the specialized approaches within social science and the humanities have not given us much understanding of its overall nature. Let us, then, examine some key elements of Newspeak:

> Newspeak words were divided into three distinct classes.... The A vocabulary consisted of words needed for the business of everyday life—for such things as eating, drinking, working, putting on one's clothes, going up and down stairs, riding in vehicles, gardening, cooking, and the like. It was composed almost entirely of words that we already possess—words like *hit, run, dog, tree, sugar, house, field*—but in comparison with the present-day English vocabulary, their number was extremely small, while their meanings were far more rigidly defined.... It would have been quite impossible to use the A vocabulary for literary purposes or for political or philosophical discussion....
>
> The grammar of Newspeak [involved] ... an almost complete interchangeability between different parts of speech. Any word in the language ... could be used either as verb, noun, adjective, or adverb.... The word *thought,* for example, did not exist in Newspeak. Its place was taken by *think,* which did duty for both noun and verb.... There was, for example, no such word as *cut,* its meaning being sufficiently covered by the noun-verb *knife.* Adjectives were formed by adding the suffix *-ful* to the noun-verb, and adverbs by adding *-wise.* Thus, for example, *speedful* meant "rapid" and

> *speedwise* meant "quickly." ... In addition, any word ... could be negatived by adding the affix *un-*, or could be strengthened by the affix *plus-*, or, for still greater emphasis *doubleplus-*. Thus, for example, *uncold* meant "warm," while *pluscold* and *doublepluscold* meant, respectively, "very cold" and "superlatively cold".... By such methods it was found possible to bring about to bring about an enormous diminution of vocabulary. Given, for instance, the word *good,* there was no need for such a word as *bad,* since the required meaning was equally well—indeed, better—expressed by *ungood*....
>
> The B vocabulary consisted of words which had been deliberately constructed for political purposes.... The B words were in all cases compound words ... the word *goodthink,* meaning, very roughly, "orthodoxy," or, if one chose to regard it as a verb, "to think in an orthodox manner." This inflected as follows: noun-verb, *goodthink*; past tense and past participle, *goodthinked*; present participle, *goodthinking*; adjective, *goodthinkful*; adverb, *goodthinkwise*; verbal noun, *goodthinker*.... Countless ... words such as *honor, justice, morality, internationalism, democracy, science,* and *religion* had simply ceased to exist.... All words grouping themselves round the concepts of liberty and equality ... were contained in the single word *crimethink,* while all words grouping themselves round the concepts of objectivity and rationalism were contained in the single word *oldthink*....
>
> The C vocabulary ... consisted entirely of scientific and technical terms.... They followed the same grammatical rules as the words in the other two vocabularies. Very few of the C words had any currency either in everyday speech or in political speech. Any scientific worker or technician could find all the words he needed in the list devoted to his own specialty, but he seldom had more than a smattering of the words occurring in the other lists ... there was no vocabulary expressing the function of Science as a habit of mind, or a method of thought, irrespective of its particular branches. There was, indeed, no word for "Science." ... (Orwell, 1949: 247–251, 254)

A fundamental characteristic of Newspeak is its focus on specialization along with a structure that works to limit communication among individuals, paralleling the present specialization throughout the social sciences with limited communication among specialists. We might recall here that there are no less than 46 Sections of the ASA. For example, the words used in the A vocabulary—emphasizing the concrete phenomena experienced in everyday life—are too narrowly and rigidly defined to be used "for literary purposes or for political or philosophical discussion." The same is true for the C vocabulary that focuses on biophysical science and technology, for those words can be used only rarely in everyday or political speech. Still more limiting is the subspecialization within the biophysical sciences, where "any scientific worker or technician could find all the words he needed in the list devoted to his own specialty, but he seldom had more than a smattering of the words occurring in the other lists." If we assume the complexity of human behavior, then such specialization and subspecialization works against our understanding human behavior and human problems. Figure 9-1 points in quite the opposite direction, for those concepts are all closely linked to one another, and that linkage is absolutely essential for penetrating human complexity.

Another basic characteristic of Newspeak is a structure that opposes the individual's ability to think, feel, and act, that is, the individual's "head," "heart," and "hand." In other words, it sets up what Jules Henry called *Culture Against Man* (1963). For example, its tiny and decreasing vocabulary together with its concepts that are narrowly and rigidly defined work against human thought, just as Postman argues that our mass media generally work in much the same direction. Further, the lack of genuine opposites takes away from the emotional force that they foster, as does the use of the affixes *plus-* and *doubleplus-* instead of qualifiers suggesting emotional force. And using the same word for both nouns and verbs removes to a large extent an individual's sense of acting decisively within any given situation. If we turn to the B vocabulary that has to do with the political sphere, we find further examples of culture against man, examples that also reflect the authoritarian structure or extreme social stratification of Oceania. The substitution of *crimethink* for "liberty" and "equality" points toward a stratified worldview and away from an evolutionary one, yet I must add that sociologists generally fail to use the concept of "cultural values," given their emphasis on social organization. The elimination of the concepts of "democracy," "science," "religion," "justice," "morality," "internationalism," and "honor" also point the individual and society in a stratified direction.

If we turn to the three major potentials that language opens up to us—dichotomous, gradational, and metaphorical or figurative thought—once again we find that Newspeak closes down human possibilities. As for dichotomies, the B vocabulary works to prevent the individual from thinking as an individual. Instead, it teaches the individual to conform to the dictates of Big Brother. Thus, only one kind of dichotomy is recognized: that between Big Brother's ideas and opposing ideas, which are labeled *crimethink* or *oldthink*, and are thus to be avoided on pain of death. As for the dichotomous concepts in Figure 9-1—concepts that can also be used gradationally and metaphorically—they are all eliminated. With respect to metaphorical or figurative thought, there is little or no room for this given the emphasis on narrow definitions of every word. Thus, the basic contribution of the humanities is lost, as we can also see from the translation of the literature of the past—that of Chaucer, Shakespeare, Milton, and Byron—into Newspeak. Those "translations" resulted in not merely changing Oldspeak into another language but changing the original works "into something different, but actually changed into something contradictory of what they used to be." Gradational thought is retained in the biophysical sciences and their technologies, for those fields could not function otherwise, but it is largely eliminated in ordinary speech and is completely eliminated in the

B language of political discourse. Overall, then, Newspeak carries much further our present limitations on the dichotomous, gradational, and metaphorical potentials of language.

Our focus on Newspeak in examining *Nineteen Eighty-Four* points up our own movement away from the infinite potentials of language for the dichotomous, gradational, and metaphorical development or evolution of the individual and society. We can see this in the parallel between our own movement toward specialization in social science and the lack of overlap among the lists of technical terms in the C language used by biophysical scientists and technologists in Oceania. We can see the lack of fulfilling language's potentials in our being captured by the trivialities flaunted by the mass media—such as the dumbing down of public debates—instead of confronting our own deepest problems, paralleling Newspeak's elimination of language accompanied by its narrowing of human consciousness. We can see the failure of most of us to become educated about the contributions of the humanities to an understanding of the human potential versus our contemporary failures, which parallels Newspeak's B language of conformity to a narrow political agenda that degrades the human being. We can also see the castration of language's potential in our society's bureaucratic specialization and stratification, which keeps everyone's mind locked into a water-tight compartment, just as Newspeak's A language is so specific and narrow that it cannot express political or philosophical thought and is thus separated from the B and C languages.

Hypothetical Ideas

Just as we began the "stereotypical ideas" section with a selection from science fiction—Orwell's *Nineteen Eighty-Four*—let us do the same for this "hypothetical ideas" section. In common with Orwell's novel, Jack Vance's *The Languages of Pao* (1958; see also Phillips, 2001: 210–214) focuses on the incredible power of language to shape human behavior, something that social science has indeed taught us this past century. More specifically, Vance's book is based on the linguistic relativity hypothesis of two anthropologists, Benjamin Lee Whorf and Edward Sapir (Whorf, 1963). Their hypothesis is that language causes people to understand and even see the world in a certain way as illustrated by three different words for three different kinds of snow in Eskimo language: snow that falls, snow shaped into steps or igloos, and snow that has accumulated on the ground. The power of language to shape our perception and understanding has been emphasized in many of the foregoing chapters as well as this one. The power of the extraordinary language of social science, as illustrated in Figure 9-1 helps us see, understand, and move toward the evolution of the individual and society. The foregoing section analyzed Newspeak and its ability to point the people of Oceania in the direction of an increasingly stratified worldview.

Unfortunately, fiction has little status among social scientists, yet its metaphors generally are unmatched throughout the literature of social science for their power to shape our understanding of human behavior.

The Languages of Pao begins in the far future on the planet of Pao, originally colonized from Earth and with fifteen billion inhabitants at the time it becomes threatened by two neighboring planets. The warlike Brumbo clan of the planet Batmarsh—emphasizing the "hand"—invade Pao successfully and then demand a huge yearly tribute from the peaceful Paonese. The rulers of the planet Mercantile—with their narrow materialistic focus on "heart"—are refusing further trade with Pao unless they too are paid an enormous sum, with Pao's relying on buying Mercantile's weapons to defend itself against the Brumbos. The scientific wizards of the planet Breakness—emphasizing the "head"—pose a potential threat for the future, given that their knowledge is vastly superior to the knowledge of the Paonese. Beran Panasper, the young son of the assassinated ruler of Pao, escapes to Breakness with the wizard, Palafox, who devises a way for the Paonese to defend themselves against the Brumbos of Batmarsh, to learn to manufacture weapons superior even to those produced by Mercantile, and to develop scientific knowledge that will prove more than a match for the knowledge of the wizards of Breakness.

Palafox reveals his plan to alter the mental framework of a portion of the Paonese people. The Paonese on three of Pao's continents would be relocated to make way for children who would learn the languages of Valiant ("hand"), Technicant ("heart"), and Cogitant ("head"). Over time, those children would, as a result, develop into warriors, industrialists, and scientists superior to those on Batmarsh, Mercantile, and Breakness, respectively. Palafox then states his explanation:

> Paonese is a passive, dispassionate language. It presents the world in two dimensions, without tension or contrast. A people speaking Paonese, theoretically, ought to be docile, passive, without strong personality development—in fact, exactly as the Paonese people are. The new language [Valiant] will be based on the contrast and comparison of strength, with a grammar simple and direct. To illustrate, consider the sentence, "The farmer chops down a tree" (literally, in Paonese, "Farmer *in state of exertion;* axe *agency;* tree *in state of subjection to attack*"). In the new language the sentence becomes: ... "The farmer vanquishes the tree, using the weapon-instrument of the axe." ... The syllabary will be rich in effort-producing gutturals and hard vowels. A number of key ideas will be synonymous; such as *pleasure* and *overcoming a resistance—relaxation* and *shame—out-worlder* and *rival.* Even the clans of Batmarsh will seem mild compared to the future Paonese military....
>
> Another area might be set aside for the inculcation of another language [Technicant].... In this instance, the grammar will be extravagantly complicated but altogether consistent and logical. The vocables would be

> discrete but joined and fitted by elaborate rules of accordance. What is the result? When a group of people, impregnated with these stimuli, are presented with supplies and facilities, industrial development is inevitable. And should you plan to seek ex-planetary markets, a corps of salesmen and traders might be advisable. Theirs would be a symmetrical language with emphatic number-parsing, elaborate honorifics to teach hypocrisy, a vocabulary rich in homophones to facilitate ambiguity, a syntax of reflection, reinforcement and alternation to emphasize the analogous interchange of human affairs.... To the military segment, a "successful man" will be synonymous with "winner of a fierce contest." To the industrialists it will mean "efficient fabricator." (Vance, 1958: 58–59)

With respect to Cogitant, the language of the scientist, it would be similar to the language of Breakness, emphasizing the "head," with almost nothing—like the range of human emotions—relating to the "heart." Beran had spent years on Breakness and recalled his experiences there:

> The language included no negativity; instead there were numerous polarities such as "go" and "stay." There was no passive voice—every verbal idea was self-contained: "to strike," "to receive-impact." The language was rich in words for intellectual manipulation, but almost totally deficient in descriptives of various emotional states ... as "anger," "joy," "love," "grief," were absent from the Breakness vocabulary. On the other hand, there were words to define a hundred different types of ratiocination, subtleties unknown to the Paonese.... On Pao there was small distinction between the sexes; both wore similar garments and enjoyed identical privileges. Here the differences were emphasized. Men wore dark suits of close-fitting fabric and black skull-caps with pointed bills. Those women whom Beran had glimpsed wore flouncing skirts of gay colors—the only color to be seen on Breakness—tight vests which left the midriff uncovered, slippers tinkling with bells. Their heads were uncovered, their hair was artfully dressed; all were young.... As he stood, a group of boys several years older than himself came up the road from the Institute; they swerved up the hill, marching in a solemn line.... Curious! thought Beran. How unsmiling and silent they seemed. Paonese lads would have been skipping and skylarking. He found his way back to Palafox's manse, puzzling over the lack of social intercourse on Breakness. (Vance, 1958: 62–63, 42–43)

Beran also recalled his learning, together with a group of Paonese boys, the languages of Cogitant, Technicant, and Valiant, for Palafox needed interpreters who would be able to communicate with everyone on Pao. On their own, apart from their studies, they had proceeded in a jocular vein to invent a patchwork tongue made up of bits and pieces of all three languages, and they christened it "Pastiche." The result of Palafox's experiment proves to be totally successful. After a single generation the Valiants develop their own warrior traditions and outfight the clansmen of Batmarsh. The Technicants build their own trading vessels, and their entrepreneurs become wildly successful at the expense of the traders on Mercantile. And the Paonese Cogitants come to rival the Wizards of Breakness. Beran makes full use of their achievements to fulfill his right to rule Pao and depose the authoritarian Bustamonte, the ruler who had tried and failed to kill him. The Cogitants, in particular, help him to oppose successfully Palafox, who had planned to make the planet of Pao his own personal territory and create his own authoritarian regime.

Yet Beran faces a final test in his efforts to help the fifteen billion inhabitants of Pao. The Valiants, Technicants, and Cogitants had very nearly replaced any loyalty to the Paonese people as a whole with fierce loyalties to their own groups. Since none of the groups are able to speak the languages of the others, none are in a position to rule the Paonese people as a whole. Given this situation, it is only Beran and his interpreters who have the necessary linguistic skills to communicate with everyone; and the Valiants, Technicants, and Cogitants finally learn their own limitations. Beran dissolves the separate enclaves that these groups had formed, dispersing them throughout the population of Pao and giving them the task of helping him and his interpreters teach everyone on the planet Pastiche, the language that includes elements of all three specialized languages. He sees this process as instilling in everyone valor, entrepreneurial ability, and wisdom. He remarks at the very end of the novel: "Pastiche—the language of service. In twenty years, everyone will speak Pastiche. It will fertilize the old minds, shape the new minds. What kind of world will Pao be then? They [Beran and his friend, Finisterle] looked out into the night, across the lights of Eiljanre, and wondered." (Vance, 1958: 159)

Newspeak and Pastiche are directly opposed to one another in many ways. The most general way is that Newspeak closes down human possibilities, pushing people further in the direction of an authoritarian or stratified and bureaucratic society along with a corresponding worldview. That language works to keep specialists separated from one another and people in general from making good use of the achievements of the specialists. Words expressing emotions are eliminated, and people learn to avoid thinking. By contrast, Pastiche opens up our possibilities for developing intellectually, emotionally, and the effectiveness of our actions: with respect to "head," "heart," and "hand." Such learning is suggested by the languages of Cogitant, Technicant (largely limited to materialistic aspirations or cultural values, although these would be extended to broader cultural values by the Paonese people), and Valiant. This broader language is illustrated in this book by the extraordinary language of social science in Figure 9-1, designed to use all three potentials of language: social science's dichotomous orientation, biophysical science's gradational orientation, and the humanities' metaphorical emphasis.

Beran's breaking up of the separate enclaves of the Cogitants, Technicants, and Valiants speaks directly to the problems that the social sciences and the rest of our world face at this time in history. How do social scientists proceed to build bridges to connect their separate worlds so that they can come to understand the complexity of human behavior? How can peoples who are presently at war with one another come to see that they have a great deal in common and that they can help one another fulfill their human potential? Beran, as the ruler of Pao, was in a position to order the breakup of those separate enclaves, to order Cogitants, Technicants, and Valiants to learn Pastiche, and to help teach Pastiche to the general population of Pao. But how might we proceed to accomplish this in actuality rather than in fiction? Further, Palafox was willing to wait a generation for those separate languages to take hold. Beran was willing to wait another generation for Pastiche to shape the behavior of the Paonese. But suppose that we cannot afford to wait for the next generation and, more realistically, that the next generation at age 20 will be in no position to challenge those who will be the leaders of society at that time. Is there any way of speeding up this process?

One limitation of Vance's vision of achieving fundamental changes in society by changing the language that people use is that it fails to pay attention to the process of perception in particular and biological structures in general.

We must focus on perceptions from one moment to the next, and not just language, if we wish to move from our semi-Newspeak language to a language closer to Pastiche. This approach will be developed and illustrated in chapter 11 where we will come closer to the practical question that is basic to this book: How can we proceed, chapter and verse, to shift from a stratified to an evolutionary worldview? Without taking perception into account, we have been ignoring our nature as animals that are the product of billions of years of biological evolution. We are, in that way, failing to build on much of who we actually are. Instead, we are focusing only on what we've become recently.

Let us assume that we cannot require any linguistic change in society as a whole, as Baran did. Let us also assume that we don't have time to wait for several generations—the 20-year-olds probably would have to wait at least a generation before they could move into positions of power—given the urgency and escalation of social problems throughout the world. Further, let us assume that it is essential to take into account both momentary perception along with the complex nature of language, just as sketched within Figure 9-1. How might we then proceed to shift our worldview along with our usage of language to open up to the incredible complexity of human behavior and human problems? Following the optimism of the scientific method, the incredible potential of language, and my own optimism as to the unlimited potential of every human being, I believe that we do indeed have the potential for moving in this direction within a time period that responds to the present urgency of the human situation. Our direction has to do with the question of motivation. If we cannot command individuals to change on the basis of conformity to authority—as did Beran—then we must find a way to motivate them to want to do so.

What is involved here are not the narrow emotional commitments of specialists like the Cogitants, Technicants, and Valiants but a broad commitment to one's own development or evolution as well as that of society. The commitment is not a narrow one to the progress of a highly specialized field of knowledge but, rather, a broad commitment to knowledge and the scientific method in general. If we return to our examination of the approach of Charles Peirce—the founder of pragmatism—in chapter 4, we find his emphasis on the importance of the scientist's motivation at the beginning of any research project. We would do well to repeat the essential idea in his argument: "The irritation of doubt is the only immediate motive for the struggle to attain belief.... With the doubt, therefore, the struggle begins, and with the cessation of doubt it ends.... There must be a real and living doubt, and without this all discussion is idle" (Peirce, 1877/1955: 10-11).

Peirce emphasizes the motivation of the scientist by returning to the first and most important step of a broad approach to the scientific method: becoming aware of and defining a problem. Without sufficient commitment or motivation, it is difficult for the scientist to stay with a problem over many years and continue to make progress. Peirce carries his motivational idea one step further when he discusses a key rule for the scientific process in general: "Do not block the way of inquiry" (Peirce, 1896/1955: 54). Inquiry should continue indefinitely, for we can always continue to make progress on a problem provided that it is a fundamental or broad one. This idea can be clarified by our pendulum metaphor: moving further in the direction of the awareness and definition of a problem can provide the momentum for moving further toward understanding the problem. The latter movement can in turn yield still further momentum toward returning to the awareness and definition of the problem, and so on.

These ideas of Peirce partly illustrate *A Theory of Cognitive Dissonance* in social psychology developed by Leon Festinger and others (1957; see also Harman-Jones et al., 1996). "Dissonance" refers to a feeling of tension or arousal produced by inconsistent thoughts, feelings, or actions, just as Peirce refers to "the irritation of doubt." Following the research of Festinger and others, individuals experiencing cognitive dissonance will attempt to relieve the tension produced by their dissonance by means of their behavior. Much of the more recent research on cognitive dissonance centers on the degree to which the individual's self-image is linked to beliefs and feelings of dissonance. Another example of cognitive dissonance is awareness of one's aspirations-fulfillment gap, as depicted in Figure I-2 and discussed in the previous chapters. Of course, individuals may proceed to bury any awareness of that gap's existence and thus attempt to avoid any consciousness of cognitive dissonance, just as the Springdalers did with their techniques of particularization and falsification of memory, as discussed in chapter 1. Yet doing so would amount to giving up on an opportunity to confront the cognitive dissonance and work toward resolving it.

A series of social psychological studies within the general framework of the theory of cognitive dissonance was inaugurated by Milton Rokeach, who focused on the idea of "self-confrontation." Rokeach brought together three institutionalized individuals—each of whom believed that he was Jesus Christ—for a period of two years (1964). In that way, he created not just a momentary situation that confronted each individual with a contradiction to his self-image but an ongoing social structure. Later, in the 1960s, 1970s, and 1980s, Rokeach and his colleagues developed a series of experiments involving situations where individuals were confronted with contradictions to their basic cultural values, such as "equality" and "freedom" (Greenstein, 1989; see also Rokeach's study of values, 1973). Rokeach's focus was on his effort to understand the causes of changes in the individual. He was interested not merely in temporary change within a given situation but also in long-term or structural change. That interest included not just beliefs and values but also actions, that is, "head," "heart," and "hand." Greenstein has summarized much of Rokeach's work:

> In the initial studies, a group of Michigan State University (MSU) students were first asked to rank the eighteen values of the Value Survey (Rokeach, 1967; see also Rokeach, 1968) in order of their "importance to you, as guiding principles in your life." Next, the students were shown results of a previous study examining the value rankings of MSU students. Then, the experimenter offered an interpretation of the value rankings of this comparison group.... The experimenter pointed out that the students in the comparison group ranked *freedom* first out of eighteen values, but ranked *equality* eleventh. At this point the experimenter attempted to arouse a state of self-dissatisfaction among the subjects by observing that the students apparently "value freedom far more highly than they value equality." This suggests that MSU students in general are much more interested in their own freedom than they are in freedom for other people....
>
> Students in the experimental groups increased their ranking for equality an average of 1.47 to 1.68 ranks three months after the experimental session.... A second set of studies was then conducted, commencing in the fall of 1967. These studies included posttest measures not only of values and attitudes but of behavior relating to civil rights as well.... Three to five months after the experimental session, all subjects received a solicitation (on NAACP letterhead) to join the National Association for the Advancement of Colored People. Experimental subjects were more than twice as likely as control subjects to respond favorably to this solicitation. Experimental subjects were significantly more likely to enroll in a special ethnic core program of coursework twenty-one months after the experimental session.... By 1979 Rokeach counted 23 published studies using the self-confrontation technique to induce long-term change. (Greenstein, 1989: 400–402; see also Phillips and Johnston, 2007: 220–223)

Rokeach, like many sociologists, was interested in long-term or structural change, including changes in cultural values and self-image, two of the central concepts in Figure 9-1. Rokeach constructed experimental situations designed to yield such change. As a result, he was able to carry the theory of cognitive dissonance further, demonstrating its links to changes in the behavior of the individual. Thus, he was able to swing the pendulum of the scientific method further in the direction of achieving understanding of just how structural change occurs. In the final chapter I shall attempt to swing that pendulum still further, centering on how an individual can achieve structural change on his or her own, without becoming the subject of such experiments. My focus will be on absolutely fundamental change, involving not only one's cultural values and self-image but also one's world-view or metaphysical stance together with the full range of structural and situational concepts depicted in Figure 9-1. Just as illustrated by Rokeach, both situations and structures will be taken into account. Situational analysis must encompass patterns of perception no less than linguistic and cultural factors.

CHAPTER 11

"Hand"

Conformity Versus Praxis

READERS SHOULD NOTE THAT TWO CRUCIAL IDEAS WILL BE INTRODUCED in this chapter: "deep dialogue" and the "East-West strategy" for confronting the problems of everyday life. Those ideas were briefly previewed in the introduction and were developed further in *Confronting Fundamental Social Problems.* Both ideas help us extend our usage of the scientific method to everyday situations from one moment to the next. The insights of the educator Paulo Freire on dialogue are the basis for my development of the concept of deep dialogue, which brings in the importance of social science knowledge to the conversational situation. "Deep dialogue" is defined in the glossary as "an egalitarian conversation where participants follow scientific ideals, reflexively uncover hidden assumptions, and move toward praxis." The East-West strategy is an effort to make full use of my discussion of Eastern thought in most of the preceding chapters. Given the emphasis on Western ideas throughout the modern world—even including the Far East—an East-West strategy can help achieve more balance in our lives and work against our growing aspirations-fulfillment gap. In the glossary an East-West strategy is defined as "an individual's approach to everyday or world problems by lowering aspirations until they are close to fulfillments (East) and then continuing to raise both aspirations and fulfillments (West)."

Just as in chapters 9 and 10, we shall begin with concepts centering on the situation. The first section of this chapter emphasizes the concept of "conformity" as linked to a stratified worldview. The second section focuses on "praxis," an element of an evolutionary worldview. These situational concepts can help us understand how structures like worldviews are constructed and how they might be changed. Much has been written that bears directly or indirectly on how one might use the scientific method in everyday life. The final section of this chapter will point in this direction and will emphasize the integration of key ideas presented earlier. For example, it will bear on questions such as: How are the ideas of "head," "heart," and "hand" linked to a broad or evolutionary approach to the scientific method? Just how do situational occurrences work to create structures? How does perception tie in to the development of structures?

Yet these questions should not lead us away from awareness of the book's title, "Armageddon or Evolution?" There is little question in my mind that time is running out for the human race. We have experienced many basic failures. For example, we have failed to become aware of how far scientific practices depart from scientific ideals, of the nature of our worldview, of our growing aspirations-fulfillment gap, and of the mammoth social and personal problems fostered by that gap. In other words, Figure I-2—which includes all of these phenomena—remains invisible to our eyes. And so do Figures 4-1, 6-1, and 9-1, for they have to do with the extraordinary language of social science, by contrast with our everyday language. What makes matters far worse is that social scientists themselves—who are responsible for the task of learning the nature of human behavior and human problems—have developed interests that are far too specialized to confront the broad phenomena portrayed in those figures. As a result, we have learned to bury awareness of the social problems portrayed in the media or our personal problems with such procedures as techniques of particularization and the falsification of memory. Just as in the case of the Springdalers—described in chapter 1—this "self-deception" and "self-avoidance" does often succeed in that it "permits some degree of satisfaction, recognition and achievement" in our day-to-day lives. Yet in the long run it is a formula for catastrophe.

We humans may still have time to create another scenario, however, and I will briefly depict my own vision of that direction in the final section, "The Scientific Method in Everyday Life." Just as in the case of global warming, we may have only a small window of opportunity, and it may already be too late. Yet we can hope that, following Thornton Wilder's play, we can somehow still manage to escape from a catastrophic future by "the skin of our teeth." Just as what occurred in that play, awareness of mounting threats to our continuing survival can motivate us to confront our fundamental problems rather than continue to hide from them. As human beings with language, we have infinite potential to learn the nature of our problems, problems that we ourselves have created. We also have the potential to act decisively and move toward solving our problems. What is involved, however, is nothing less than our most fundamental assumptions, our metaphysical stance, our worldview. For that worldview points us toward a narrow scientific method that prevents us from reaching more than a superficial understanding of our situation.

Conformity

Let us look to the extraordinary language depicted in Figure 9-1 (page 167) to locate "conformity"—or conforming behavior within a given momentary situation—in relation to personality and social structures. Orwell's *Nineteen Eighty-Four*—discussed in chapter 10—describes many situations where individuals conform to the dictates of Big Brother, such as everyone's watching the Two Minutes of Hate on the telescreens and expressing the required hate against Big Brother's enemies. In Orwell's final paragraph his hero, Winston Smith, finally sobs after a lifetime of failed efforts to rebel against Big Brother's ruthless tyrannical regime:

> He gazed up at the enormous face. Forty years it had taken him to learn what kind of smile was hidden beneath the dark mustache. O cruel, needless misunderstanding! O stubborn, self-willed exile from the loving breast! Two gin-scented tears trickled down the sides of his nose. But it was all right, everything was all right, the struggle was finished. He had won the victory over himself. He loved Big Brother. (1949: 245)

Winston's attempts at "deviance" from the dictates of Big Brother took place against a background of conformity in a great many situations, such as his attendance at the sessions of the Two Minutes of Hate, and his doing what was required by his superiors while working at the Ministry of Truth—for example, revising history so that it conformed to the requirements of the Party. But the real triumph of Big Brother is illustrated in the quoted paragraph: altering Winston's beliefs and values, with reference to Big Brother, from hate to love. This is achieved not just by the patterns of social organization within Oceania, as exemplified by the torture Winston experienced in Room 101. Let us not forget that the culture of Oceania—as illustrated by the language of Newspeak—and Oceania's patterns of social organization pushed its inhabitants toward conformity. In other words, it was Oceania's social structure as a whole, not simply the actions of Big Brother and his henchmen. As a result, there is enormous pressure toward conformity—on pain of death—with respect to thinking, feeling, and acting, or "head," "heart," and "hand." Winston must be taught not merely to act in accordance with Big Brother's wishes, he must learn to love Big Brother, just as those watching their two-way telescreens must exhibit appropriate emotions. The change in Winston's personality structure has further consequences. For that change, in turn, influences Winston to conform willingly to the many requirements of Big Brother.

Yet all of this is fiction. What about real life? For example, can the individual be induced to conform to acts that go against his or her fundamental values and beliefs? Among the many studies on this subject, the Milgram series of experiments on obedience to authority stand out in their clear demonstration of just how far the individual can be influenced to conform without being forced to conform, as in the case of the residents of Big Brother's Oceania. Stanley Milgram, a psychologist at Yale University, secured the participation of 500 subjects for his experiments by advertising in a local paper as well as by calling people randomly from a telephone directory, and paying for services rendered. Milgram described his experiment as having to do with the process of teaching and learning. A subject would play the role of the "teacher"; the "learner" would be someone who—unknown to the participants—had been carefully coached on how to behave. He was a 47-year-old man whom subjects later described as "mild-mannered and likable."

The experiment called for the learner to remember pairs of words and to be given an electric shock each time he failed to remember. The "shock generator" was, unknown to the subjects, actually a bogus instrument, with voltages supposedly ranging from 15 to 450. Teachers were required by the experimenter—a stern-looking individual in a gray technician's coat—to increase the voltage in increments of 15 volts for each additional incorrect answer. Milgram set up four different experimental conditions, ranging from a remote condition where the learner was in an adjoining room to a situation where the teacher was required to place the learner's hand on the shock plate. As for the results, 65 percent of the teachers continued to shock the learner up to the maximum 450 volts, with the percentage declining to 62.5 percent, 40 percent, and 30 percent in the three conditions representing greater degrees of proximity between teacher and learner. Teachers continued to shock the learner with what they knew were potentially lethal voltages while the learner screamed in agony.

The following remarks by a teacher in the remote condition illustrate the contradiction between his words and his actions of continuing to shock the learner: ". . . these are terrific volts. I don't think this is very humane. . . . Oh, I can't go on with this; no, this isn't right. It's a hell of an experiment. The guy is suffering in there. No, I don't want to go on. This is crazy" (Milgram, 1974: 32). Milgram described his experiment to audiences, while withholding information about the results he obtained. He posed a question: Would the teachers refuse to give potentially lethal shocks? Of the 110 people in those audiences, all responded that the teachers would refuse to go past 300 volts. Most of them indicated that no one would go past 150 volts. How are we to explain the great disparity between the predictions of those audience members and the actual results Milgram obtained? Why did most of those predictions indicate that the teachers would rebel at a relatively low 150 volts? And why did every respondent in those audiences conclude that no one would continue past 300 volts?

Despite the vast numbers of social science studies indicating the existence of powerful patterns of social stratification throughout contemporary society, those audience members appear to have remained ignorant of such findings. Similarly, they appear to have been ignorant of a stratified worldview operating behind the scenes of the experiment, a system of fundamental assumptions that supports hierarchical behavior in all areas of life throughout contemporary society. Those concepts and findings alert us to the power of authority figures—such as professors like Yale University's Milgram as well as technicians in gray coats—to influence our behavior, given the credibility that they have. This widespread ignorance of the findings of social scientists extends far beyond this lack of understanding of the power of social stratification in modern society. It even extends to social scientists themselves, as illustrated by the minimal communication among the specialists in ASA's 46 Sections, let alone by communication with individuals in other social sciences.

Yet there appears to be more to the explanation of the gap between the audience members' predictions and Milgram's experimental results. If we recall the discussion of cultural values in chapter 6, people-oriented values included individual personality and equality, which would suggest concern for the health of the learner in Milgram's experiment. Such people-oriented values then would presumably influence the audience members to predict limitations on the severity of the electric shocks that the teachers administered. Overall, the discrepancy between the audience members' predictions and the experimental results is another example of the wide aspirations-fulfillment gap throughout contemporary society. Our cultural values call for egalitarian behavior that respects the ultimate worth of the individual. Yet the experimental results indicated that actual behavior departs very far from those values. We are surprised when alerted to this gap, for we are accustomed to burying any awareness of it with our techniques of particularization and our falsification of memory.

Given the power of social stratification—and social organization in general—in modern society, what about the power of culture to induce conformity? In particular, we might recall the importance of work-related cultural values—as discussed in chapter 6—especially those of "material comfort" and economic "progress." We might see our materialistic civilization as built largely on the basis of the incredible successes of biophysical science technologies, giving us the ability to shape our physical environment over the past five centuries as never before in human history. Yet it is our one-sided emphasis on materialistic values—given the failure of social science to develop far enough to yield the basis for effective social technologies—that is closely linked to the full range of our social problems, to our stratified worldview, and to our patterns of conformity.

Let us review here Schumacher's indictment of our one-sided materialistic economy (see chapter 8):

> What is the meaning of democracy, freedom, human dignity, standard of living, self-realisation, fulfillment? Is it a matter of goods, or of people? Of course it is a matter of people.... If economic thinking ... cannot get beyond its vast abstractions, the national income, the rate of growth, capital/output ratio, input-output analysis, labour mobility, capital accumulation; if it cannot get beyond all this and make contact with the human realities of poverty, frustration, alienation, despair, breakdown, crime, escapism, stress, congestion, ugliness, and spiritual death, then let us scrap economics and start afresh. (Schumacher, 1973: 70)

Although the breadth of Buddhist economics contrasts sharply with Western economics, as Schumacher explains (see p. 123), nevertheless he claims: "The choice of Buddhism is purely incidental; the teachings of Islam, or Judaism could have been used as well as those of any other of the great Eastern traditions" (1973: 49).

A sign that I recently saw in front of a church read: "The best things in life are not things." This is a well-intentioned effort to oppose modern materialism, yet it is a feeble effort compared to the multibillion-dollar advertising campaigns throughout the media that point in the opposite direction. Advertisements are becoming ever more ubiquitous. Few among us would confess to being influenced by advertising, given that such conformity goes against cultural values like freedom, equality, and individual personality. Yet why would corporations spend billions on advertising if that spending did not produce results? Advertising is the key to the furthering of an economy that depends on mass sales. That economy is intimately linked to the full range of social structures throughout modern society—as depicted in Figure 9-1—social structures that are in turn linked to all of our personality structures. Advertising is linked to much more than contemporary culture, as Figure 9-1 suggests. Following Schumacher's indictment of our materialistic economy, advertising is also linked to such problems as "poverty, frustration, alienation, despair, breakdown, crime, escapism, stress, congestion, ugliness, and spiritual death."

Looking once again to Figure 9-1, it is time to shift this analysis from "conformity" to "praxis," a word based on the Greek root "prak," meaning "to effect," "to do." Similar words depend on this Greek root such as "practical," "practice," "pragmatic," and the philosophy of "pragmatism." It is that philosophy more than any other that is the basis for the broad approach to the scientific method that I adopt throughout this book. Let us recall from chapter 4 some of the ideas of Charles Peirce, the founder of pragmatism: "The irritation of doubt is the only immediate motive for the struggle to attain belief.... With the doubt, therefore, the struggle begins, and with the cessation of doubt it ends" (1877/1955: 10). Just as in Nietzsche's *The Gay Science* (1887/1974; see chapter 3), Peirce makes emotions central to the scientific method. Let us also recall Peirce's rule "which itself deserves to be inscribed upon every wall of the city of philosophy: Do not block the way of inquiry" (1877/1955: 10). Science, from this perspective, becomes an evolutionary process; unfortunately, contemporary social scientists have yet to discover how to follow this fundamental rule of the scientific method. We should not ignore, however, Peirce's view of the centrality of metaphysics for the scientific method and his lament about the present state of this body of knowledge (see chapter 7): "There is and can be no doubt that this immature condition of Metaphysics has very greatly hampered the progress of ... psychology ... linguistics, anthropology, social science, etc. To my mind it is equally clear that defective and bad metaphysics has been almost as injurious to the physical sciences" (1898/1955: 310). I am hopeful that my own approach to metaphysics throughout this book and earlier ones, contrasting a stratified worldview or metaphysical stance with an evolutionary worldview, will provide a better basis for the scientific method.

Praxis

We might begin by seeing "praxis" as action that fulfills a wide range of values. Just like "conformity," "deviance," and "social interaction," "praxis" is momentary action within a given situation, by contrast with physical, biological, personality, and social structures. This distinction is crucial for understanding both how situational behavior can, over time, yield structures, and also just how structures can yield behavior within a given scene. My approach here is quite similar to that of Anthony Giddens. Giddens's theory of "structuration" is oriented to give attention to both situational actions by individuals—or, more generally, "agents"—and social structures, although his many theoretical contributions extend considerably beyond that idea (see, for example, Giddens, 1984; www.lse.ac.uk/Giddens.publications.htm). If we look at Figure 9-1 we can see that the definition of praxis includes both the values of the individual as well as values that are widely shared, such as cultural values, thus invoking both individual structures and social structures. If we look at the right side of Figure 9-1, we see that praxis is linked to both rituals and social rituals. Although praxis is situational action, when it persists over time it can yield rituals and social rituals that are very meaningful and expressive. By contrast, the action of conformity—at least within a stratified and bureaucratic society or group—would over time yield rituals that are far less meaningful and expressive. As for the aspirations-fulfillment gap of the individual as well as anomie throughout society, praxis points toward narrowing that gap and reducing anomie, given that it "fulfills a wide range of values."

The concept of praxis is fundamental to the educational theories and practices of Paulo Freire, a Brazilian educator who taught literacy to Brazil's peasants, often achieving significant results in as little as 30 hours of instruction. Freire was appointed director of the Brazilian National Literacy Program in 1963, but he was jailed after a military coup in 1964. He then went into exile, only to return in 1979 to found the Workers Party. His major work, *Pedagogy of the Oppressed* (1970), can help us understand not only the nature of praxis but also a range of the concepts in Figure 9-1. Freire's focus is on how to change social structures to solve social problems, and he illustrates a Western orientation to praxis.

We shall follow this brief analysis by taking up an Eastern orientation to changing personality structures and solving certain individual problems. We shall examine *The Twenty Guiding Principles of Karate,* as developed by Gichin Funakoshi and written up by Genwa Nakasone (2003). These guiding principles do not emphasize the specific moves within patterns of fighting but rather karate's Buddhist ideas and values, thus illustrating praxis as well as other concepts within Figure 9-1. Combining these two analyses, what emerges is an East-West orientation to praxis and an evolutionary worldview that meshes with earlier treatments in this book of Hindu, Buddhist, and Confucian ideas.

Central to Freire's thought is what he sees as the basic theme of modern society:

> I consider the fundamental theme of our epoch to be that of *domination*—which implies its opposite, the theme of *liberation,* as the objective to be achieved.... While the problem of humanization has always ... been man's central problem, it now takes on the character of an inescapable concern.... The current movements of rebellion, especially those of youth ... place consumer civilization in judgment, denounce bureaucracies of all types, demand the transformation of the universities ... propose the transformation of reality itself so that universities can be renewed.... The central problem is this: How can the oppressed, as divided, unauthentic beings, participate in developing the pedagogy of their liberation? Liberation is thus a childbirth, and a painful one. The man who emerges is a new man, viable only as the oppressor-oppressed contradiction is superseded by the humanization of all men. (1970: 93, 27, 33)

Freire first published this book in Portuguese in 1968, during the height of the countercultural movement in the United States and elsewhere. His language of "domination" and "liberation" adds a good deal to the rather bland concepts in Figure 9-1, such as "social stratification." His language illustrates a deep sense of problem, just as Mills's language illustrated. That sense of problem is better represented in Figures I-1, I-2, and I-3 than in Figure 9-1. His problem of achieving "humanization" and "liberation" is much the same as the fundamental problems presented in those figures: How do we change from a stratified to an evolutionary worldview and from an increasing to a decreasing aspirations-fulfillment gap, with both changes working to reduce a range of threatening social and personal problems?

As an educator, Freire has much more to say about how to achieve humanization and liberation, based on his own experiences. For example, he focuses more specifically on the nature of the problem of domination:

> banking education ... mirror[s] oppressive society as a whole: (a) the teacher teaches and the students are taught; (b) the teacher knows everything and the students know nothing; (c) the teacher thinks and the students are thought about ... (h) the teacher chooses the program content, and the students (who were not consulted) adapt to it.... Self-depreciation is another characteristic of the oppressed, which derives from their internalization of the opinion the oppressors hold of them. So often do they hear that they are good for nothing, know nothing and are incapable of

> learning anything—that they are sick, lazy, and unproductive—that in the end they become convinced of their own unfitness.... One of the characteristics of oppressive cultural action which is almost never perceived by the dedicated but naive professionals who are involved is the emphasis on a *focalized* view of problems rather than on seeing them as dimensions of a *totality.* In "community development" projects the more a region or area is broken down into "local communities," without the study of these communities both as totalities in themselves and as parts of another totality (the area, region, and so forth)—which in its turn is part of a still larger totality (the nation, as part of the continental totality)—the more alienation is intensified. (1970: 49, 59, 137-138)

Freire specifies other problems as well. For one thing, he sees domination and oppression not just as a matter of social stratification but also as a matter of cultural oppression. Thus, his analysis stretches over both culture and social organization in Figure 9-1. To illustrate, there is the "focalized" or narrow view of "dedicated but naive professionals." We might think here of social scientists in general and sociologists in particular, recalling the ASA with its 46 distinct Sections and little communication among their members. This example shows the bureaucratic aspects of a stratified or bureaucratic worldview, given the emphasis of bureaucracy on specialization with limited communication. Yet, given the enormous complexity of human behavior, it becomes essential to build bridges connecting all of these specialized areas—and reaching out to the other social sciences and the humanities as well—to gain a sense of the "totality" of human behavior. We might view such specialization with limited communication as illustrative of the "cultural repression" of students and of their instructors as well. And we might view Freire's examples of "banking education" as exemplifying repression by means of social organization. As for the "self-depreciation ... of the oppressed," who are all of us, we might come to see this as a product of both kinds of domination. Questions have been raised as to whether Freire's analysis applies to first-world as well as third-world countries; it appears that his work applies to contemporary society in general.

Granting that praxis involves a recognition of fundamental problems in society and oneself, it also calls for action to confront those problems effectively. We should remember that Freire was extremely effective in educating peasants and that he wrote during a time of widespread and fundamental problems in society—the counterculture era—when many were optimistic about the possibilities of achieving fundamental changes. How can the individual proceed, through praxis and dialogue or "social interaction," to transform self and society so as to move toward liberation or humanization?

> As we attempt to analyze dialogue as a human phenomenon, we discover something which is the essence of dialogue itself: *the word.* But the word

> is more than just an instrument which makes dialogue possible; accordingly, we must seek its constitutive elements. Within the word we find two dimensions, reflection and action, in such radical interaction that if one is sacrificed—even in part—the other immediately suffers. There is no true word that is not at the same time a praxis. Thus, to speak a true word is to transform the world.... When a word is deprived of its dimension of action, reflection automatically suffers as well; and the word is changed into idle chatter, into *verbalism,* into an alienated and alienating "blah." It becomes an empty word, one which cannot denounce the world, for denunciation is impossible without a commitment to transform, and there is no transformation without action. On the other hand, if action is emphasized exclusively to the detriment of reflection, the word is converted into *activism.* The latter—action for action's sake—negates the true praxis and makes dialogue impossible. (1970: 75-76)

To understand Freire's argument here, let us substitute "useful word" for "true word," given that the social sciences have indeed produced a great many "true words"; but by failing to integrate them so as to confront human complexity—and by failing to confront their own patterns of social stratification—they have also failed to produce few "useful words." From this perspective, Freire's critique cuts to the bone of the social sciences at this time in history. Social scientists' continuing to cling to their ivory towers, largely avoiding responsibility for what is now happening in the world, convert their many truths into "idle chatter, into *verbalism,* into an alienated and alienating 'blah.'" As for those activists throughout the world who focus on "action for action's sake" while avoiding "reflection," their *activism* "negates the true praxis and makes dialogue impossible." We might see such activists as including a variety of different groups, all of whom proceed—and have proceeded in the past—in attempting to change the world with little understanding of what is involved. For example, there were the students and others within the countercultural movement of the 1960s who believed that they already were in possession of the truth and required no additional knowledge of human behavior, thus negating the many achievements—granting their lack of integration—of the social sciences. There are the contemporary political figures in the West who hold the reins of power—as well as many of the talking heads in the mass media—who believe, similarly, that they are already in possession of the truth and can safely ignore whatever the social sciences have discovered.

Freire goes on to examine the nature of genuine or authentic dialogues that point toward egalitarian "social interaction," humanization, and liberation:

> Dialogue is the encounter between men, mediated by the world, in order to name the world.... If it is in speaking their word that men, by naming the world, transform it, dialogue imposes itself as the way by which men

> achieve significance as men. Dialogue is thus an existential necessity ... dialogue cannot be reduced to the act of one person's "depositing" ideas in another, nor can it become a simple exchange of ideas to be "consumed" by the discussants. Nor yet is it a hostile, polemical argument between men who are committed neither to the naming of the world, nor to the search for truth, but rather to the imposition of their own truth ... it is [rather] conquest of the world for the liberation of men.... Dialogue cannot exist, however, in the absence of a profound love for the world and for men.... Domination reveals the pathology of love: sadism in the dominator and masochism in the dominated.... On the other hand, dialogue cannot exist without humility. The naming of the world, through which men constantly re-create that world, cannot be an act of arrogance.... Dialogue further requires an intense faith in man, faith in his power to make and remake, to create and re-create, faith in his vocation to be more fully human (which is not the privilege of an elite, but the birthright of all men).... Nor yet can dialogue exist without hope. Hope is rooted in men's incompletion, from which they move out in constant search—a search which can be carried out only in communion with other men. Hopelessness is a form of silence, denying the world and fleeing from it. (Freire, 1970: 76–80)

Freire's vision of dialogue is most egalitarian, by no means abandoning the cultural achievements of the West with their emphasis on the cultural values of equality, freedom, democracy, and individual personality, as discussed in chapter 6. Yet that vision has no place for the West's—and the East's—development of patterns of social stratification and bureaucracy that work to trash those values. Further, Freire's focus is on the small group and the individual, given his personal involvement as a educator of peasant groups. Thus, his concern is with large-scale and small-scale social structures, and also with personality structures. Following Figure 9-1, his overriding emphasis is on the situational act of praxis. For him, "naming the world" is not the empty *verbalism* to be found in the academic world, where so-called dialogues are "consumed" by the participants with no movement toward action to solve problems and change the world. Such *consumption* of ideas is closely linked to the West's materialistic way of life along with our advertising juggernaut that has become a driving force for that way of life. These so-called dialogues—that we encounter so often in today's world—are "hostile, polemical argument between men who are committed neither to the naming of the world, nor to the search for truth, but rather to the imposition of their own truth" (1970). Freire links the kind of dialogue that embodies praxis to the emotions of "love," "humility," "faith," and "hope." For him, praxis not only includes "head" and "hand" but also "heart." This is a most scientific stance if we are to follow the ideas of Peirce on the importance of "the irritation of doubt" (chapter 4) and Nietzsche's idea of "gay science" (chapter 3).

Freire's vision of the incredible power of language to "name the world" and, by so doing, to transform the world—by contrast with using words for the purpose of "consumption"—may sound like a strange idea. How can language and dialogue possibly change the world? Yet let us bear in mind that language—more than anything else—distinguishes us human beings from all other forms of life. Language is our most powerful tool for solving problems. Also, language gave us the basis for carrying forward that power by developing the scientific method. And language gives us the basis for continuing to develop our "head," "heart," and "hand" without any limit.

If we turn to the glossary after this chapter, we can begin to give credibility to Freire's idea of using language to transform the world, which is the basis for his concept of praxis. Let's note, for example, how "individual" is defined: "A human being with the potential for continually developing "head," "heart," and "hand." That definition, then, points toward changing all individuals and, by so doing, changing the world. For another example, we might look to the definition of "language": "A system of concepts, widely shared throughout society, for solving problems by means of its dichotomous, gradational, and metaphorical potentials for understanding and communicating experience." Yet my argument throughout this book is that the individual has not traveled very far down the road of using those three potentials; thus, the definition once again points toward changing the world. Still another example has to do with the six definitions for the economic, educational, family, political, religious, and scientific institutions. Those definitions portray institutions as centering on solving a given problem. My argument throughout the book, indirect as well as direct, is that they have largely failed us in their efforts to solve those problems. Thus, those definitions hold out to us the problems that are yet to be solved and, as a result, give us a direction for changing the world.

My approach to defining dialogue in the glossary is to build on Freire's ideal that dialogue should be egalitarian, that it should be a search for truth, and that it should yield praxis. I call such dialogue "deep dialogue," with this definition: "An egalitarian conversation where participants follow scientific ideals, reflexively uncover hidden assumptions, and move toward praxis." I have added the idea that participants in a deep dialogue "reflexively uncover hidden assumptions." This constitutes deep learning on their part, for such assumptions have to do with our worldview. If our worldview is not uncovered, we continue to remain prisoners of it from one moment to the next.

Let us now turn from the West to the East, building on ideas from Hinduism, Buddhism—including those of Gandhi—and Confucianism that we have already examined in chapters 2-8. Many of us are familiar with Rudyard Kipling's "The Ballad of East and West"—as quoted in chapter 4—which includes this stanza:

> Oh, East is East, and West is West, and never the twain shall meet,
> Till Earth and Sky stand presently at God's great Judgement Seat:
> But there is neither East nor West, Border, nor Breed, nor Birth,

> When two strong men stand face to face, tho' they come from the ends of earth!

Generally, we may not think of Kipling as an advocate of East-West interaction, paying attention to the first two lines of the ballad, but the second two lines suggest the possibilities of such interaction. I believe that this interaction is absolutely vital at this time in history. As I've sketched in chapters 2 and 3, it is an East-West synthesis that provides a strategy and tactics for moving toward an evolutionary worldview.

Gichin Funakoshi was the father of modern karate—one of the martial arts—on the island of Okinawa at the beginning of the twentieth century. In 1922 he was invited to Tokyo by the Ministry of Education to demonstrate this new martial art, and he stayed on there in an effort to popularize it. While teaching karate at universities, at police headquarters, and in other places, Funakoshi proceeded to study Zen Buddhism. As a result he developed further the spiritual or humanistic aspects of karate-do. "Do" means "way," as in "Bushido," or the "Way of the Warrior"; "karate" means "empty hand," for those who follow karate-do—"the Way of the Empty Hand"—have no weapons other than their own bodies. The possession of weapons had been banned in Okinawa, and karate had been developed among warriors as a substitute for weapons and as a matter of self-defense. Our purpose here is by no means to discuss the specific moves or "kamae" and "kata" associated with karate-do but rather to illustrate the nature of praxis and an evolutionary worldview from an Eastern perspective. Our focus here will be on seven of *The Twenty Guiding Principles of Karate*—retaining their original numbers—although all of them are closely related to an evolutionary worldview.

> *1. Do not forget that Karate-do begins and ends with* Rei. . . . *Rei* is often defined as "respect," but it actually means much more. *Rei* encompasses both an attitude of respect for others and a sense of self-esteem. When those who honor themselves transfer that feeling of esteem—that is, respect—to others, their action is nothing less than an expression of *rei*. . . . It should also be noted that although a person's deportment may be correct, without a sincere and reverent heart they do not possess true *rei*. True *rei* is the outward expression of a respectful heart. (Funakoshi and Nakasone, 2003: 19–20)

We can draw a parallel here between *rei* and Gandhi's techniques of civil disobedience, as discussed in chapter 8. We might recall that Gandhi treated Lord Irwin, the British viceroy, and the British Army with great respect. For example, Gandhi refrained from any form of verbal or physical violence toward them and taught his followers to do the same. He told them of his plans, refrained from acting to humiliate them, initiated personal interaction with them, adopted an attitude of trust toward them, and attempted to achieve a high degree of empathy with respect to their values and beliefs. Looking to Figure 9-1, *rei* thus has to do, not just with the individual's momentary behavior, but also with personality and small-scale social structures. The personality structures of values, beliefs, and rituals must become involved, focusing on egalitarian behavior. This implies, on a larger scale, the importance of cultural values along with social rituals that are meaningful and expressive. More generally, Gandhi's behavior, along with *rei,* suggests movement toward an evolutionary worldview.

> *3. Karate stands on the side of justice....* Human beings are at their strongest when they believe they are right. The strength that comes from the confidence of someone who knows he or she is right is expressed by the saying, "When I examine myself and see that I am in the right, then whether I am faced by one thousand or ten thousand opponents, I must press onward." To avoid action when justice is at stake demonstrates a lack of courage. (2003: 29–30)

Funakoshi sees karate-do as involving nothing less than the entire individual, for the integration of the individual—including "head," "heart," and "hand"—is essential for the development of the "confidence" that is essential for success in this martial art. The evolutionary worldview and all of the concepts in Figure 9-1 are oriented toward helping the individual learn to integrate "head," "heart," and "hand" along with biological and physical structures. There is also the idea of self-sacrifice involved in this orientation to justice, for it requires the individual to confront any number of opponents simultaneously if he or she determines that this is the correct course of action. We might think here of Hitler's subordinates within Nazi Germany who claimed that they had to follow orders to kill millions of civilians during the Holocaust, or they themselves would have been killed. We might also consider Gandhi's orientation to making visible sacrifices for his cause, such as his long periods of fasting that carried him to the point of near death.

> *4. First know yourself, then know others....* Karate practitioners must be completely aware of their own strengths and weaknesses, and never become dazzled or blinded by conceit or overconfidence. Then they will be able to assess calmly and carefully the strengths and weaknesses of their adversaries, and create an ideal strategy. (2003: 33–34)

Funakoshi is referring to the importance of personality structures no less than social structures, as depicted in Figure 9-1. More specifically, our stratified and bureaucratic worldview points us outward in our perceptions within everyday life, by contrast with the inward-outward orientation of an evolutionary worldview. We have seen the importance of the latter orientation in relation to Levin's experiment, as described in chapter 4. There, the self-evaluators with their inward-outward orientation did not increase their levels of prejudice against Puerto Ricans upon being frustrated. More generally, it is that inward-outward orientation to perception that can work to place us—figuratively—on a stairway pointing toward our continuing evolution. We shall see the importance of this in the final section, "The Scientific Method in Everyday Life."

> *6. The mind must be set free* ... our mind should be allowed to move about freely, even if it seeks muddy recesses. The lotus blossom is not sullied by the mud in which it grows.... To reign in the mind tightly takes away its freedom. To keep our mind in close confines may be a necessary beginner's habit, but doing so for our entire life prevents us from rising to a new level, and will result in a life of unfulfilled potential. (2003: 43, 45)

An emphasis on "conformity"—whether to the wishes of others or to cultural norms or widely shared expectations—chains the mind to narrow tasks. Indeed, there are many ways in which we chain the mind so that we fail to make use of more than an infinitesimal fraction of the mind's full potential. For example, our stratified worldview keeps us locked into one narrow compartment of human experience or another. As another general example, we all tend to emphasize language's dichotomous potential (emphasized within social science) in our everyday thought and speech, while we largely ignore language's gradational potential (emphasized within biophysical science) and language's figurative or metaphorical potential (emphasized within the humanities). Figure 9-1 suggests the possibilities that language holds out to us for learning to use more of all three linguistic potentials as we go about our everyday lives.

> *8. Karate goes beyond the dojo.* The objective of karate-do is to polish and nurture both the mind and body. The cultivation of one's spirit and mental attitude begun during practice in the *dojo* (training hall) should not cease after the physical and mental exertions end for the day. Rather, this should continue outside the *dojo,* in our daily routine. (2003: 53)

For there to be a change from situational to structural behavior, behavior must be repeated, and that is exactly what the eighth principle of karate-do calls for. This transition is particularly important for martial arts, since the individual must respond with lightning speed to be effective. Thus, there is little time to think about what to do when an opponent makes a move. Once a principle becomes part of an individual's personality structure, however, it becomes available in an instant.

> *9. Karate is a lifelong pursuit.* There is no single point that marks the completion of karate training; there is always a higher level. For this reason practitioners should continue training throughout their life.... Walking this endless road, becoming better today than yesterday, and then better

> tomorrow than today—throughout one's life—is a true image of the Way of Karate. (2003: 57, 59)

This figure or metaphor of an endless road applies to an evolutionary worldview as well as to the scientific method. Another useful figure—that can be applied to karate-do, to an evolutionary worldview, and also to the scientific method—is that of a pendulum swinging in ever-widening arcs. To one side there is increasing awareness of problems, and to the other side, ever more progress toward understanding and solving them. But since every problem is linked to every other problem and since we are becoming increasingly aware of problems, we are never in a situation of solving any problem completely.

> *16. When you step beyond your own gate, you face a million enemies.* Negligence is a great enemy when we leave the safety of our homes. If we are not in peak form in both our body and attitude, we will attract troublemakers and problems. Consequently, we should adopt the attitude that when leaving our gate we are entering into the midst of many potential enemies and should stay mentally alert. (2003)

The title of this book—"Armageddon or Evolution?"—reveals my own sense of the magnitude of world and personal problems. Those problems are everywhere, including inside our gates as well as outside them, and—following Figure I-1—they are increasing. Those problems are illustrated in Figure 9-1 by such concepts as alienation, aspirations-fulfillment gap, addiction, negative reinforcement, relative deprivation, conformity, anomie, social stratification, and bureaucracy. Yet we can learn to change our negative to positive reinforcements, adopting an East-West strategy of lowering our aspirations in the short run so as to narrow the gap (Eastern orientation), and thus learning to integrate "head," "heart," and "hand." We can then learn to follow this by raising both aspirations and their fulfillment in the long run (Western orientation).

The Scientific Method in Everyday Life

"Armageddon or Evolution?" suggests my own belief that fundamental problems in modern society—like the relatively invisible aspirations-fulfillment gap and its links to the visible problems such as dangers from small groups with weapons of mass destruction—are escalating, and that a continuation of these developments even threatens the continuing existence of the human race. What is equally dangerous, in my view, is that political leaders along with their experts appear to be largely ignorant of the scope of these problems, their nature, or how to confront them effectively. Yet the book's title also suggests the possibility that these problems can be solved. The subtitle, "The Scientific Method in Everyday Life," more specifically points toward a direction for solving them. Figure I-3 is still more specific in charting that direction, where a broad scientific method can yield the change from a stratified to an evolutionary worldview. That change in turn can narrow the aspirations-fulfillment gap (Figure I-1), with the result of enabling us to confront social and personal problems ever more effectively.

I might roughly categorize my argument throughout the book into seven areas, which I briefly examine in the remaining pages: (1) Evolution-do Versus Stratification-do or Bureaucracy-do, (2) Perception and an East-West Strategy, (3) "Heart," (4) "Head," (5) "Hand," (6) From Situational Behavior to Structures, and (7) Some Concluding Remarks. As I write these words, I'm aware that I am accentuating the positive more than the threatening aspects of modern society's present situation, granting the deepening problems that exist and the lack of any political direction that I see for meeting those threats effectively. Yet I do believe that optimism is called for in these times, provided that it is based on realism. My experiences lead me to believe that the directions I have put forward in this book and summarize here are in fact realistic.

(1) Evolution-do Versus Stratification-do or Bureaucracy-do

Throughout this book I have attempted to shed light on the question of how we can all move from our stratified and bureaucratic worldview to an evolutionary and interactive one. The "do," taken from Japanese, indicates the "way" or a way of life, and these contrasting worldviews are indeed contrasting ways of life. "Evolution-do" is much the same as "the Way of the Empty Hand" in that the individual must take personal responsibility for changing self and world, by contrast with relying on leaders to do the job for him or her, just as warriors not practicing karate-do rely on weapons in their hands. Evolution-do is illustrated in this final chapter by the principles of karate-do summarized previously as well as by those principles not summarized here. For example, the first principle, "Do not forget that Karate-do begins and ends with *Rei,*" points up the importance of heartfelt egalitarian relationships between opponents or, more generally, between any two individuals. This is exactly Freire's approach to dialogue, for such egalitarian interaction points toward liberation. The third principle of karate-do, "Karate stands on the side of justice," suggests that this martial art should stand for value fulfillment by contrast with alienation, anomie, and a one-sided emphasis on materialism.

(2) Perception and an East-West Strategy

The East-West strategy I've put forward is to lower one's aspirations in the short run so as to narrow the aspirations-fulfillment gap and, as a result, shift away from negative reinforcements toward positive reinforcements, thus following a Buddhist orientation. Given that narrowed gap, the individual can then learn, in the long run, to continue to raise both aspirations and their fulfillment, following a Western orientation. This orientation also follows the ideals of Confucius, who saw "the path of duty lies in what is near," while "men seek for it in what is remote" (in Kaplan, 1961: 276). This approach meshes with the fourth karate-do principle, "First know yourself, then know others." For being able to perceive one's own weaknesses within a given situation would lead to the lowering of one's aspirations to a more realistic level, thus enabling them to be fulfilled more easily. More generally, by emphasizing one's perception of the realistic possibilities within any given situation—by contrast with inattention to the momentary scene—one can proceed to do the same.

(3) "Heart"

Commitment to addressing a problem is not only the first step of a broad approach to the scientific method, it is by far the most important one. More generally, the same commitment to the problems encountered in everyday life, both personal and world problems, is crucial for moving toward solutions. Unfortunately, given our stratified worldview, we have all learned to repress such problems to varying degrees. That worldview has stood in the way of understanding the nature of fundamental problems—by professional social scientists along with everyone else—and we thus have no effective direction for confronting them. Yet with the aid of the extraordinary language of social science, as illustrated in Figure 9-1, we move into a position to confront our problems effectively. Following (2) presented previously, Narrowing the aspirations-fulfillment gap can help us express our emotions and become deeply committed to solving our problems.

(4) "Head"

Movement toward an evolutionary worldview will help us use language's dichotomous, gradational, and metaphorical or figurative potentials. That movement will teach us the importance of learning social science's "extraordinary language," as illustrated in Figure 9-1. That language can help us integrate our ordinary language to an increasing extent, and that integration can in turn provide the basis for improving our ability to express emotions and commit to problems ("heart"). All of this adds up to penetrating the complexity of current fundamental social problems, by contrast with the present relative helplessness of our leaders. It also adds up to a vision of "man-the-scientist," by contrast with our present view of the ordinary individual as unable to do more than support specialized scientists in their work. It is a vision that includes the idea that one day we will all learn to be poets, philosophers, and scientists.

(5) "Hand"

For Freire, language that does not achieve theoretical reflection is no basis for genuine dialogue, liberation, or humanization. Rather, it meshes with empty activism, just as Funakoshi saw an approach to karate without an emphasis on spirituality in the same way. Similarly, words that are merely consumed and lead nowhere become mere verbalism that goes nowhere. By contrast, language can be used to "name the world." Thus, words can and should be translated into actions that change the world. This is the nature of praxis. Yet much more than dialogue is involved within praxis, such as love, humility, faith, and hope. Just as social structures generally have shaped individuals, it is extremely important for the reverse to occur to a much greater extent: the shaping of social structures by individuals. Indeed, that is essential if we are to move from a stratified to an evolutionary worldview.

(6) From Situational Behavior to Structures

Funakoshi's principles 8, 9, and 6 all have to do with the transition from situational behavior to the development of structures. For the latter requires the repetition of behavior in one situation after another. Such repetition is fostered in principle 8 when the student continues with his or her training in everyday life outside the training hall. It is further fostered by following principle 9, when karate-do becomes a lifelong pursuit. As for principle 6, once the individual develops structures and no longer needs to focus the mind narrowly on following the principles of karate-do, he or she is able to relax and free the mind. All of this is similar to learning to use social science's extraordinary language—as illustrated by Figure 9-1—in everyday life. Initially, one learns to focus the mind narrowly on learning to use those concepts in more situations. Then, over time, the individual alters his or her linguistic structures and is then free to relax the mind.

(7) Some Concluding Remarks

The end of our journey in this book is also the beginning of the next journey. For this book is no more than the *dojo* or training hall for evolution-do. Following principles 8 and 9, moving toward the structures required by evolution-do involves both practice beyond the *dojo* and practice as a lifelong pursuit. The development of the structures of evolution-do, following principle 6, then frees the mind and the relaxed mind will continue to evolve without limit. That continuing evolution suggests the operation of the pendulum metaphor. Commitment to problems ("heart") gives the pendulum motivation to swing further in the direction of understanding ("head") and solving ("hand") those problems. For example, an individual can learn to address the problem of bringing his or her negative emotions to the surface ("heart"), then learn to gain momentum by coming to reinforce self positively for doing so, and thus achieve greater understanding of those emotions ("head") and greater ability to express self ("hand"). The extraordinary language of social science—invoking perception in the momentary situation and sensitivity to physical, personality, and social structures—would necessarily be involved in "head" within this sequence for the individual to penetrate the complexity of any given momentary scene. "Hand" would have to succeed in converting negative to positive reinforcements. As a result, those "head" and "hand" achievements would yield greater momentum in bringing emotions to the surface in other situations, leading over time to the development of personality structures for doing so. Unfortunately, "Armageddon or Evolution?" still remains a question, for there is no guarantee that social scientists and the rest of us will move through our small window of opportunity and learn to confront our escalating problems more effectively.

I do not want to end this book on a passive note, however. We human beings have infinite potential to evolve intellectually, emotionally, and in the effectiveness of our actions. Our most powerful tools—language and the scientific method—give us the potential to solve our most difficult problems, including our present stratified or bureaucratic worldview and scientific method. That potential can carry our development ever further, far beyond present-day problems. The ideas of deep dialogue and the East-West strategy provide a direction for using our linguistic and scientific tools from one moment to the next in our everyday lives. In my view, at stake is not just those of us who are now struggling with increasing problems but also all future generations. All of us—social scientists and everyone else—are responsible for what happens now and what will happen in the future. If we succeed in shouldering that responsibility—based on our learning the nature of our situation, envisioning ways of solving our problems, and engaging in praxis to solve them—then there is no limit to how far we humans can evolve.

Glossary

Addiction An individual's persisting pattern of action that subordinates individuality to dependence on external phenomena

Alienation An individual's persisting feelings of isolation from self, others, one's own biological structure, and physical structures

Anomie Substantial aspirations-fulfillment gaps widely shared throughout society

Aspirations-Fulfillment Gap The individual's failure to achieve certain values

Assumptions Beliefs that may be invisible to the individual or group

Beliefs Persisting ideas or opinions of an individual or group

Biological Structure A structure that interacts to a relatively great extent with its environment

Bureaucracy A group with an extensive hierarchy and division of labor that has been deliberately constructed to solve specific problems

Conformity An individual's situational behavior that follows the requirements of group norms or values or the requirements of an individual or group

Cultural Norms The beliefs or expectations and assumptions widely shared throughout society

Cultural Values The values, interests, ideals, or goals widely shared throughout society

Cultural Worldview The worldview widely shared throughout society

Culture The values (including interests, ideals and goals) and norms (including beliefs and assumptions) of a people or a society

Deep Dialogue An egalitarian conversation where participants follow scientific ideals, reflexively uncover hidden assumptions, and move toward praxis

Definition of the Situation An individual's or group's understanding of the nature of a momentary scene

Deviance An individual's situational behavior that departs from the requirements of group norms or values or the requirements of an individual or group

East-West Strategy An individual's approach to everyday or world problems by lowering aspirations until they are close to fulfillments (East) and then continuing to raise both aspirations and fulfillments (West)

Economic Institution An institution in society centering on the problems of producing, distributing, and consuming whatever society has come to value

Educational Institution An institution in society centering on the problem of communicating an understanding of the universe, including physical, biological, social, and personality structures

Epistemology Fundamental assumptions and beliefs—that derive from a given metaphysics—about how to learn or obtain knowledge

Evolutionary Worldview A worldview emphasizing egalitarian social interaction and using the scientific method to continually develop the individual's "head," "heart," and "hand"

Expressive Orientation An individual's commitment to awareness and expression of his or her emotions

Family An institution in society centering on the problems of achieving intimate or very close social relationships as well as the continuation of life from one generation to the next

Group An aggregate of individuals who repeatedly interact directly or indirectly

Hypothetical Ideas Ideas or thoughts about the nature of phenomena based on the scientific method

Individual A human being with the potential for continually developing "head," "heart," and "hand"

Institution A system of cultural values and norms within a society focused on solving a given problem

Language A system of concepts, widely shared throughout society, for solving problems by means of its dichotomous, gradational, and metaphorical potentials for understanding and communicating experience

Metaphysics Fundamental assumptions and beliefs about the nature of reality that shape all human behavior

Negative Reinforcement The lack of fulfillment of an individual's values within a momentary situation

Norms The persisting beliefs or expectations and assumptions of a group

Perception An individual's sensation—sight, hearing, touching, tasting, and/or smelling—in a momentary situation influencing his or her definition of the situation

Personality Structure The individual's system of values, beliefs and assumptions, and rituals

Physical Structure A structure that interacts to a relatively small extent with its environment

Political Institution An institution in society centering on the problems of developing, distributing, and using power

Positive Reinforcement The fulfillment of an individual's values within a momentary situation

Praxis An individual's action within a given momentary situation that fulfills a wide range of his or her values and moves toward solving personal and world problems

Reflexive Ideas An individual's inward-looking ideas within a given momentary situation

Relative Deprivation An individual's feelings within a momentary scene of an unjustified failure to fulfill values relative to others who are seen as enjoying greater fulfillment

Religious Institution An institution in society centering on the problem of finding ultimate meaning or significance in life

Ritual An individual's persisting pattern of action that is meaningful and expressive to some degree

Scientific Institution An institution in society centering on the problem of developing understanding of the universe, including physical, biological, social, and personality structures

Scientific Method Procedures for solving problems based on an epistemology that emphasizes (1) defining problems so as to build on relevant previous knowledge, (2) testing ideas against experience, (3) assessing the researcher's impact on the research process, and (4) continuing to increase one's understanding without any limit by endlessly repeating this process

Scientific Technology Technology based on the scientific method

Self-image An individual's view of self

Situation A location in time and space

Social Interaction Action by individuals in a momentary situation that mutually affects them

Social Organization Persisting and shared patterns of action and interaction within a group

Social Relationship Persisting social interaction between individuals

Social Ritual A ritual that is shared within a group

Social Stratification A persisting hierarchy or pattern of inequality within a group or among groups

Social Structure Shared values, beliefs and assumptions, and rituals within a group

Society A group with a culture and patterns of social organization

Stereotypical Ideas Fixed ideas—that are limited or erroneous—about the nature of phenomena

Stratified Worldview A worldview emphasizing social stratification and bureaucracy

Structure A persisting system of elements

Technology A persisting pattern of action for solving problems

Values Persisting motives, interests, ideals, or goals of an individual or group

Worldview The individual's metaphysical stance or persisting image of reality

References

Abbott, Edwin A. *Flatland: A Romance of Many Dimensions.* New York: Dover, 1884/1952.

Addams, Jane. *Democracy and Social Ethics.* New York: Macmillan, 1902.

Banks, Jane, and Jonathan David Tankel. "Science as Fiction: Technology in Prime Time Television," *Critical Studies in Mass Communication* 7 (1990), 24–36.

Bannister, D., and Fay Fransella. Inquiring Man: The Theory of Personal Constructs. New York: Penguin, 1971.

Bardach, Janusz. *Man Is Wolf to Man: Surviving the Gulag.* Berkeley, Calif.: University of California Press, 1998.

Barry, Ellen. "Study Looks at Loss, Its Role in Depression: Finds Humiliation Triggers Worst Cases," *The Boston Globe,* August 13, 2003, A3.

Baum, L. Frank. *The Wizard of Oz and the Land of Oz.* New York: Random House, 1960.

Begley, Sharon. "Older Really Is Wiser, Research Finds: Though Stereotyped as Feeble, Aging Brains Are Actually Finely Tuned," *Sarasota Herald-Tribune,* February 17, 2007, 1A, 17A.

Berger, John. *Ways of Seeing.* London: BBC and Penguin Books, 1972.

Bondurant, Joan V. *Conquest of Violence.* Berkeley, Calif.: University of California Press, 1965.

Borgatta, Edgar F., and Rhonda J. V. Montgomery. *Encyclopedia of Sociology.* New York: Macmillan Reference USA, 2000.

Bowles, Samuel, and Herbert Gintis. "IQ in the United States Class Structure," in Alan Gartner, Colin Greer, and Frank Riessman (eds.), *The New Assault on Equality: I.Q. and Social Stratification.* New York: Harper & Row, 1974.

Brown, Donald E. *Human Universals.* New York: McGraw-Hill, 1991.

Buckley, Walter. *Sociology and Modern Systems Theory.* Englewood Cliffs, N.J.: Prentice-Hall, 1967.

Buruma, Ian. "Theatre of War," *The New York Times Book Review,* September 17, 2006, 1, 9–10.

Busch, Lawrence. "Macrosocial Change in Historical Perspective: An Analysis of Epochs." Unpublished Ph.D. dissertation. Ithaca, N.Y.: Cornell University, 1974.

———. "A Tentative Guide to Constructing the Future: Self-Conscious Millenarianism," *Sociological Practice* 1 (Spring 1976), 27–39.

Carroll, James. "The Culture of Self-Deception," *The Boston Globe*, July 30, 2002, A15.

Constas, Helen. "Max Weber's Two Conceptions of Bureaucracy," *American Journal of Sociology* 52 (January 1958), 400–409.

Contemporary Sociology 35 (November 2006), 551–558.

Darwin, Charles. *The Origin of Species.* New York: New American Library, 1859/1958.

Davis, Kingsley, and Wilbert E. Moore, "Some Principles of Stratification," *American Sociological Review* 10 (April 1945), 242–249.

Denzin, Norman K. Review of Bernard Phillips's *Beyond Sociology's Tower of Babel. Contemporary Sociology* 31 (November 2002), 790–792.

Domhoff, G. William. "Mills's *The Power Elite* 50 Years Later," *Contemporary Sociology* 35 (November 2006), 547–550.

Durkheim, Emile. *Suicide.* New York: Free Press, 1897/1951.

Ellison, Ralph. *Invisible Man.* New York: Vintage, 1947.

Erikson, Kai. "On Work and Alienation," *American Sociological Review* (February 1986), 1–8.

Festinger, Leon. *A Theory of Cognitive Dissonance.* Stanford, Calif.: Stanford University Press, 1957.

Footnotes 35 (February 2007). Washington D.C.: American Sociological Association.

Freire, Paulo. *Pedagogy of the Oppressed.* New York: Herder and Herder, 1970.

Fromm, Erich. *The Sane Society* (chapter 9). New York: Rinehart, 1955.

Funakoshi, Gichin, and Genwa Nakasone. *The Twenty Guiding Principles of Karate.* Tokyo: Kodansha International Ltd., 2003.

Gartner, Alan, Colin Greer, and Frank Riessman (eds.). *The New Assault on Equality: I.Q. and Social Stratification.* New York: Harper & Row, 1974.

Gerth, Hans H., and C. Wright Mills (eds.). *From Max Weber: Essays in Sociology.* New York: Oxford University Press, 1946.

———. *Character and Social Structure: The Psychology of Social Institutions.* New York: Harcourt, Brace, and World, 1953.

Giddens, Anthony. *The Constitution of Society.* Berkeley: University of California Press, 1984.

Gould, Kenneth A., David N. Pellow, and Allan Schnaiberg. *The Treadmill of Production: Injustice and Unsustainability in the Global Economy.* Boulder, Colo.: Paradigm Publishers, 2008.

Gould, Kenneth A., Allan Schnaiberg, and A. S. Weinberg. *Local Environmental Struggles: Citizen Activism in the Treadmill of Production.* Cambridge: Cambridge University Press, 1996.

Gould, Stephen Jay. *The Mismeasure of Man.* New York: Norton, 1981.

Gouldner, Alvin W. *The Coming Crisis of Western Sociology.* New York: Basic Books, 1970.

———. "The Politics of the Mind: Reflections on Flack's Review of *The Coming Crisis of Western Sociology*," *Social Policy* 5 (March/April 1972), 13–21, 54–58.

Greenstein, Theodore N. "Modifying Beliefs and Behavior Through Self-Confrontation," *Sociological Inquiry* 59 (November 1989), 396–407.

Harman-Jones, E., J. Brehm, J. Greenberg, L. Simon, and D. E. Nelson. "Evidence That the Production of Dissonance Is not Necessary to Create Cognitive Dissonance," *Journal of Personality and Social Psychology* 70 (1996), 5–16.

Hart, Betsy. "We Live in a Can't-Say-No Culture," *Sarasota Herald-Tribune,* February 14, 2007, 13A.
Henry, Jules. Culture Against Man. New York: Random House, 1963.
Herling, Gustav. *A World Apart.* New York: Roy Publishers, 1951.
Herrick, C. Judson. *The Evolution of Human Nature.* Austin: University of Texas Press, 1956.
Hesse, Hermann. *The Glass Bead Game (Magister Ludi).* New York: Bantam Books, 1943/1969.
Horney, Karen. *The Neurotic Personality of Our Time,* New York: Norton, 1937.
Hornig, Susanna. "Television's NOVA and the Construction of Scientific Truth," *Critical Studies in Mass Communication* 7 (1990), 11–23.
Huxley, Aldous. Brave New World. New York: Harper & Row, 1939.
Illich, Ivan. *Deschooling Society.* New York: Harper & Row, 1972.
James, William. *Pragmatism: A New Name for Some Old Ways of Thinking.* New York: Dover, 1907/1995.
Janis, Irving L., and Daniel Katz. "The Reduction of Intergroup Hostility," *Journal of Conflict Resolution* 3 (March 1959), 85–100.
Jones, Alexander. "Ecclesiastes," *The Jerusalem Bible,* 855–864. Garden City, N.Y.: Doubleday, 1966.
Kaplan, Abraham. *The New World of Philosophy.* New York: Random House, 1961.
Kelly, George A. *A Theory of Personality: The Psychology of Personal Constructs.* New York: W. W. Norton, 1963.
Knowles, Elizabeth (ed.). *The Oxford Dictionary of Quotations.* New York: Oxford University Press, 2004.
Knottnerus, J. David. "Agency, Structure and Deritualization: A Comparative Investigation of Extreme Disruptions of Social Order," in Sing C. Chew and J. David Knottnerus (eds.), *Structure, Culture and History: Recent Issues in Social Theory.* Lanham, Md.: Rowman & Littlefield, 2002.
———. *Ritual as a Missing Link within Sociology: Structural Ritualization Theory and Research.* Boulder, Colo.: Paradigm Publishers, forthcoming.
Knottnerus, J. David, and Bernard Phillips (eds.). *Bureaucratic Culture and Escalating Problems.* Boulder, Colo.: Paradigm Publishers, forthcoming.
Korzybski, Alfred. *Science and Sanity.* Garden City, N.Y.: Country Life Press, 1933.
Kuhn, Thomas S. *The Structure of Scientific Revolutions.* Chicago: University of Chicago Press, 1962.
Laszlo, Ervin. *Introduction to Systems Philosophy: Toward a New Paradigm of Contemporary Thought.* New York: Harper & Row, 1973.
Lenski, Gerhard. *Ecological-Evolutionary Theory.* Boulder, Colo.: Paradigm Publishers, 2005.
Lerner, Daniel. *The Passing of Traditional Society.* New York: Free Press, 1958.
Levin, Jack. "The Influence of Social Frame of Reference for Goal Fulfillment on Social Aggression." Ph.D. dissertation. Boston: Boston University, 1968.
Lundberg, George A. *Can Science Save Us?* New York: David McKay, 1947/1961.
Marx, Karl. "Essay on Alienation," in Karl Marx, *Early Writings,* ed. and trans. by T. B. Bottomore. New York: McGraw-Hill, 1844/1964.
———. "theses on Feuerbach," in Robert C. Tucker (ed.), *The Marx-Engels Reader.* New York; W. W. Norton, 1972, 108–109.
McGraw, Phillip C. *Self Matters: Creating Your Life from the Inside Out.* New York: Simon and Schuster, 2001.

Milgram, Stanley. *Obedience to Authority.* New York: Harper & Row, 1974.

Mills, C. Wright. "Situated Actions and Vocabularies of Motive," *American Sociological Review* 5 (December 1940), 904–913.

———. "The Professional Ideology of Social Pathologists," *American Journal of Sociology* 49 (1943), 165–180.

———. *The New Men of Power: America's Labor Leaders.* New York: A. M. Kelley, 1948.

———. *White Collar: The American Middle Classes.* New York: Oxford University Press, 1951.

———. *The Power Elite.* New York: Oxford University Press, 1956.

———. *The Causes of World War Three.* New York: Simon & Schuster, 1958.

———. *The Sociological Imagination.* New York: Oxford University Press, 1959.

———. *Listen, Yankee: The Revolution in Cuba.* New York: McGraw-Hill, 1960.

———. *The Marxists.* New York: Dell Publishing, 1962.

———. *Sociology and Pragmatism: The Higher Learning in America.* New York: Oxford University Press, 1964.

Mills, Kathryn, with Pamela Mills (eds.). *C. Wright Mills: Letters and Autobiographical Writings.* Berkeley: University of California Press, 2000.

Montefiore, Simon Sebag (ed.). *Speeches That Changed the World.* London: Smith-Davies, 2005.

Murray, Donald M. "In Life as in War, Overcoming Fear One Small Step at a Time," *The Boston Globe,* October 30, 2001, E2.

Nietzsche, Friedrich. *The Gay Science,* trans. by Walter Kaufmann. New York: Vintage, 1887/1974.

Orwell, George. *Nineteen Eighty-Four.* New York: Harcourt Brace Jovanovich, 1949.

Osgood, Charles, et al. *The Measurement of Meaning.* Urbana: University of Illinois Press, 1967.

Paul, Pamela. "Proceed to Checkout," *The New York Times Book Review,* April 8, 2007, 15.

Peirce, Charles S. "The Fixation of Belief," in C. S. Peirce, *Philosophical Writings of Peirce,* 5–22. New York: Dover, 1877/1955.

———. "How to Make Our Ideas Clear," in C. S. Peirce, *Philosophical Writings of Peirce,* 23–41. New York: Dover, 1878/1955.

———. "The Scientific Attitude and Fallibilism," in C. S. Peirce, *Philosophical Writings of Peirce,* 42–59. New York: Dover, 1896/1955.

———. "The Approach to Metaphysics," in C. S. Peirce, *Philosophical Writings of Peirce,* 310–314. New York: Dover, 1898/1955.

Perrine, Laurence. *Sound and Sense: An Introduction to Poetry,* 4th ed. New York: Harcourt, 1973.

Phillips, Bernard. *Social Research: Strategy and Tactics.* New York: Macmillan, 1966, 1971, 1976.

———. *Sociology: Social Structure and Change.* New York: Macmillan, 1969.

———. *Sociology: From Concepts to Practice.* New York: McGraw-Hill, 1979.

———. *Sociological Research Methods: An Introduction.* Homewood, Ill.: Dorsey Press, 1985.

———. *Beyond Sociology's Tower of Babel: Reconstructing the Scientific Method.* New York: Aldine de Gruyter, 2001.

———. "Mills, C(harles) Wright (1916–1962)," in John R. Shook (ed.), *Dictionary of Modern American Philosophers.* Bristol, UK: Thoemmes Press, 2004.

——— (ed.). *Understanding Terrorism: Building on the Sociological Imagination.* Boulder, Colo.: Paradigm Publishers, 2007.

Phillips, Bernard, and Louis C. Johnston. *The Invisible Crisis of Contemporary Society: Reconstructing Sociology's Fundamental Assumptions.* Boulder, Colo.: Paradigm Publishers, 2007.

Phillips, Bernard, Harold Kincaid, and Thomas J. Scheff (eds.). *Toward a Sociological Imagination: Bridging Specialized Fields.* Lanham, Md.: University Press of America, 2002.

Plato. *Plato's Republic,* trans. G. M. A. Grube. Indianapolis, Ind.: Hackett, 1974.

Polak, Fred L. *The Image of the Future.* San Francisco: Jossey-Bass, 1973.

Postman, Neil. *Amusing Ourselves to Death: Public Discourse in the Age of Show Business.* New York: Penguin, 1985.

Rafalovich, Adam. "Assessing the Fallout of the Terrorist Moment: Anomie and the Fractured American *Weltanschauung*" (chapter 4), in Bernard Phillips (ed.), *Understanding Terrorism: Building on the Sociological Imagination.* Boulder, Colo.: Paradigm Publishers, 2007.

Raushenbush, Hilmar S. *Man's Past: Man's Future.* New York: Delacorte, 1969.

Rees, Martin. *Our Final Hour: A Scientist's Warning: How Terror, Error, and Environmental Disaster Threaten Humankind's Future in This Century—on Earth and Beyond.* New York: Basic Books, 2003.

Retzinger, Suzanne. *Violent Emotions: Shame and Rage in Marital Quarrels.* Newbury Park, Calif.: Sage, 1991.

Rich, Frank. *The Greatest Story Ever Sold: The Decline and Fall of Truth from 9/11 to Katrina.* New York: Penguin Press, 2006a.

———. "Yes, You Are the Person of the Year!" *The New York Times,* Sunday, December 24, 2006b, Week in Review, 8wk.

Rokeach, Milton. *The Three Christs of Ypsilanti.* New York: Knopf, 1964.

———. *Value Survey.* Sunnydale, Calif.: Halgren Tests, 1967.

———. *Beliefs, Attitudes, and Values.* San Francisco: Jossey-Bass, 1968.

———. *The Nature of Human Values.* New York: Free Press, 1973.

Sanderson, Stephen K. *Evolutionism and Its Critics: Deconstructing and Reconstructing an Evolutionary Interpretation of Human Society.* Boulder, Colo.: Paradigm Publishers, 2006.

Satter, Raphael G. "Doomsday Clock Inches a Little Closer to Midnight," *Sarasota Herald-Tribune,* January 8, 2007, 7A.

Scheff, Thomas J. *Microsociology: Discourse, Emotion, and Social Structure.* Chicago: University of Chicago Press, 1990.

———. *Bloody Revenge: Emotions, Nationalism, and War.* Boulder, Colo.: Westview, 1994.

———. *Emotions, the Social Bond, and Human Reality: Part/Whole Analysis.* Cambridge: Cambridge University Press, 1997.

———. *Goffman Unbound! A New Paradigm for Social Science.* Boulder, Colo.: Paradigm Publishers, 2006.

Schnaiberg, Allan. *The Environment: From Surplus to Scarcity.* New York: Oxford University Press, 1980.

Schnaiberg, Allan, and Kenneth A. Gould. *Environment and Society: The Enduring Conflict.* New York: St. Martin's Press, 1994.

Schumacher, E. F. *Small Is Beautiful: Economics as if People Mattered.* New York: Harper & Row, 1973.

Simmel, Georg. "Metropolis and Mental Life," in Georg Simmel, *On Individuality and Social Forms,* ed. Donald N. Levine, 227-234. Chicago: University of Chicago Press, 1903/1971.

Sommer, Robert. *Tight Spaces: Hard Architecture and How to Humanize It.* Englewood Cliffs, N.J.: Prentice-Hall, 1974.

Sperling, Laura. "Every Second Counts, or Does It?" *Sarasota Herald-Tribune,* February 17, 2007, 19A.

Sztompka, Piotr. *The Sociology of Social Change.* Cambridge, Mass.: Blackwell, 1993.

Thomas, Cal. "Rise of the Anti-Hero," *Sarasota Herald-Tribune,* February 22, 2007, 11A.

Thomas, William I. *The Unadjusted Girl.* Boston: Little, Brown, 1923.

Udy, Stanley H., Jr. "'Bureaucracy' and 'Rationality' in Weber's Organization Theory," *American Sociological Review* 24 (December 1959), 591-595.

Van Delinder, Jean. *Struggles before Brown: Early /Civil Rights Protests and Their Significance Today.* Boulder, Colo.: Paradigm Publishers, 2007.

Van Vogt, A. E. *The World of Null-A.* New York: Berkley Publishing, 1945/1970.

———. *The Players of Null-A.* New York: Berkley Publishing, 1948.

Vance, Jack. *The Languages of Pao.* New York: Daw Books, 1958.

Vidich, Arthur, and Joseph Bensman. *Small Town in Mass Society: Class, Power, and Religion in a Rural Community.* Garden City, N.Y.: Doubleday, 1960.

Von Bertalanffy, Ludwig. *General System Theory.* New York: George Braziller, 1968.

Weber, Max. *The Protestant Ethic and the Spirit of Capitalism.* New York: Scribner's, 1905/1958.

———. *The Theory of Social and Economic Organization.* New York: Free Press, 1964.

Whitman, Walt. "Song of Myself," in *Leaves of Grass,* 23-76. New York: Random House, 1892/1983.

Whorf, Benjamin. *Language, Thought, and Reality.* Cambridge, Mass.: MIT Press, 1963.

Wiener, Norbert. *The Human Use of Human Beings: Cybernetics and Society.* Garden City, N.Y.: Doubleday Anchor, 1954.

Wiesel, Elie. *U.S. News and World Report,* October 27, 1986, quoted in Elizabeth Knowles (ed.), *The Oxford Dictionary of Quotations,* 6th ed., 834.13. New York: Oxford University Press, 2004.

Williams, Robin M., Jr. *American Society,* 3rd ed., 452-500. New York: Knopf, 1970.

Wrong, Dennis. "The Functional Theory of Stratification: Some Neglected Considerations," *American Sociological Review* 24 (1959), 772-782.

Index

About the Author

Bernard Phillips, a student of C. Wright Mills at Columbia, received an M.A. from Washington State University and a Ph.D. from Cornell, studying with Robin M. Williams, Jr. Phillips taught at the University of North Carolina and the University of Illinois before teaching at Boston University. He cofounded ASA's Section on Sociological Practice and—in addition to articles—wrote a number of textbooks in research methods and introductory sociology. He founded the Sociological Imagination Group—with the help of Harold Kincaid and Thomas J. Scheff—which has had annual conferences since 2000. Phillips initiated a series of monographs together with Harold Kincaid, "Advancing the Sociological Imagination," first with Aldine and then—along with his new coeditor, J. David Knottnerus—with Paradigm Publishers. His most recent books are *Beyond Sociology's Tower of Babel,* the edited volume *Toward a Sociological Imagination* (with Kincaid and Scheff), *The Invisible Crisis of Contemporary Society* (with Louis C. Johnston), the edited volume *Understanding Terrorism,* and the edited volume with J. David Knottnerus, *Bureaucratic Culture and Escalating Problems* (in press with Paradigm). He is currently working on a manuscript with an immodest title, "How to Save the World: From a Bureaucratic to an Evolutionary Way of Life." Excerpts from his work are on the website www.sociological-imagination.org. His e-mail address is bernieflps@aol.com. He lives in Longboat Key, Florida, and Provincetown, Massachusetts, with his wife Marjorie. Their children are David and Michael.